BRITISH GEOLOGICAL SURVEY

R. A. OLD, M. G. SUMBLER
and K. AMBROSE

CONTRIBUTORS

Stratigraphy
J. Brewster
A. R. L. Jones
A. W. A. Rushton

Geophysics
R. M. Carruthers

Hydrogeology
P. K. Murti

Palaeontology
M. A. Calver, H. C. Ivimey-Cook,
N. J. Riley and G. Warrington

Petrology
R. K. Harrison and G. E. Strong

Geology of the country around Warwick

Memoir for 1:50 000 geological sheet 184
(England & Wales)

Natural Environment Research Council

LONDON: HER MAJESTY'S STATIONERY OFFICE   1987

Bibliographical reference

OLD, R. A., SUMBLER, M. G. and AMBROSE, K. 1987. Geology of the country around Warwick. *Mem. Br. Geol. Surv.*, Sheet 184 (England & Wales).

Authors

R. A. OLD, BSc, PhD
M. G. SUMBLER, BA
K. AMBROSE, BSc
British Geological Survey, Keyworth

Contributors

R. M. Carruthers, BSc, H. C. Ivimey-Cook, BSc, PhD, N. J. Riley, BSc, PhD, A. W. A. Rushton, BSc, PhD, G. E. Strong, BSc and G. Warrington, BSc, PhD
British Geological Survey, Keyworth

M. A. Calver, MA, PhD
Cock's Hill, Heveningham, Halesworth, Suffolk IP19 0EJ

R. K. Harrison, MSc
27 Springfield Park, Twyford, Berkshire

A. R. L. Jones, BSc
40 Upper Packington Road, Ashby-de-la-Zouch, Leicester LE6 5EF

P. K. Murti, MSc
formerly British Geological Survey, Wallingford

J. Brewster, BSc
Marius Skadsens Vei 29, N4 300, Sandnes, Norway

Other publications of the Survey dealing with this district and adjoining districts

BOOKS

Memoirs
Geology of the country around Market Harborough, Sheet 170
Geology of the country around Stratford upon Avon and Evesham, Sheet 200
Geology of the country around Banbury and Edge Hill, Sheet 201

British Regional Geology
Central England (3rd edition)

MAPS

1:1 000 000
Pre-Permian geology

1:625 000
Solid geology (South sheet)
Quaternary geology (South sheet)
Aeromagnetic map (South sheet)

1:250 000
Solid geology, East Midlands
Aeromagnetic anomaly, East Midlands
Bouguer gravity anomaly, East Midlands

1:63 360 and 1:50 000
Sheet 168 Birmingham, Solid, Drift
Sheet 169 Coventry, Solid, Drift
Sheet 170 Market Harborough, Solid and Drift
Sheet 183 Redditch, Solid and Drift (in press)
Sheet 184 Warwick, Solid and Drift
Sheet 185 Northampton, Solid and Drift
Sheet 200 Stratford upon Avon, Solid and Drift
Sheet 201 Banbury, Solid and Drift
Sheet 202 Towcester, Solid and Drift

Printed in the United Kingdom for Her Majesty's Stationery Office
Dd 240406 C20 9/87 398 12521

Geology of the country around Warwick

The district described in this memoir incorporates parts of the Coventry Horst and the Hinckley and Knowle basins, important post-Triassic structures which probably had pre-Triassic origins.

The oldest rocks, of Upper Cambrian age, are entirely concealed by younger strata, but have been proved in numerous deep boreholes, and broad stratigraphic and structural units have been identified. Similarly the Carboniferous below the Coventry Sandstone has been proved by boreholes, sunk mainly in search of coal, and below the Keele Formation the beds are known in more detail. The district includes the northern part of the South Warwickshire Prospect, a major extension of the Warwickshire Coalfield syncline south of Coventry. A description is given of the Thick Coal, the main seam of economic importance. The Halesowen Formation includes coal seams which are correlated with those of the Oxfordshire Coalfield.

Above the Halesowen Formation the Upper Coal Measures are red, poorly fossiliferous, continental deposits. Rocks of the Enville Group, at outcrop in the north west, span the Carboniferous – Permian boundary.

Continuing continental sedimentation until the close of Triassic times, was controlled partly by subsidence on the margins of the Knowle and Hinckley basins. A major marine transgression in Penarth Group times led to the establishment of marine conditions which persisted throughout the deposition of most of the highly fossiliferous Jurassic sequence.

Stratified glacial drift deposits, the product of the Wolstonian glaciation, include the type area of this stage. Later Quaternary deposits include extensive river terraces.

Sections of the memoir deal with the tectonic history, mineral products and hydrogeology of the district.

Plate 1 (Frontispiece). Warwick Castle from the east. The castle is constructed from locally quarried Bromsgrove Sandstone Formation.   A13092

CONTENTS

PLATES

TABLES

NOTES

Throughout the memoir the word 'district' refers to the area covered by the Warwick (184) Sheet.

National Grid references are given in square brackets. Unless otherwise stated all lie within the 100 km square SP.

Thin sections prefixed by 'E' are those in the Survey's collection of English sliced rocks.

PREFACE

This memoir describes the geology of the district covered by the Warwick (184) Sheet 1:50 000 Geological Map of England and Wales. The district was originally surveyed on the scale of one inch to one mile by W. T. Aveline, H. H. Howell, A. C. Ramsay and R. Trench and the results published between 1854 and 1859 on Old Series Sheets 53NW, 53NE, 53SW, 53SE, 54NE and 54SE. Small areas along the northern, eastern and southern margins were surveyed on the scale of six inches to one mile during the present century, as overlap from the mapping of the adjoining one-inch sheets 169 (Coventry), 185 (Northampton) and 201 (Banbury).

The whole of the district was surveyed on the 1:10 000 scale by Messrs K. Ambrose, J. Brewster and M. G. Sumbler and Dr R. A. Old between 1976 and 1978. The survey was under the supervision of Mr G. W. Green as District Geologist. The 1:10 000 maps covered by the various surveyors are listed on p.viii. The solid with drift edition of the 1:50 000 map was published in 1984.

The writing and compilation of this memoir has been shared mainly between Messrs Ambrose and Sumbler and Dr Old. Dr A. W. A. Rushton contributed largely to the Cambrian chapter. The chapters on igneous rock, the Westphalian and Permian, the Triassic, economic geology and structure were written by Dr Old; the Jurassic chapter by Mr Ambrose and the Quaternary chapter by Mr Sumbler. The account of geophysical investigations was written by Mr R. M. Carruthers and the section on water supply by Mr P. K. Murti. Dr M. A. Calver advised on the Westphalian stratigraphy and Dr N. J. Riley identified macrofossils from the Westphalian. The Triassic microfossils were identified by Dr G. Warrington, and the Jurassic macrofossils by Dr H. C. Ivimey-Cook. Mr R. K. Harrison and Mr G. E. Strong supplied petrological data for several chapters. The memoir was edited by Mr E. A. Edmonds.

The Rugby Portland Cement Company supplied much of the information about the history of cement manufacture. Grateful acknowledgement is made to Mr A. R. L. Jones and the staff of the National Coal Board, South Midlands Area, for valuable advice, for permission to publish data from the concealed Warwickshire Coalfield, and for providing a contribution on this area.

G. Innes Lumsden, FRSE
Director

British Geological Survey
Keyworth
Nottingham NG12 5GG

2 July 1987

LIST OF 1:10 000 MAPS

Geological National Grid 1:10 000 maps included wholly or in part in Sheet 184 are listed below together with the initials of the surveyors and the dates of the survey.

The surveyors were: K. Ambrose, J. Brewster, R. A. Old and M. G. Sumbler for Sheet 184 and C. H. Cunnington, E. A. Edmonds, R. H. Hoare, E. G. Poole, P. J. Strange, B. J. Williams and V. Wilson for the peripheral areas. All the maps are available as photographic or dyeline prints. Those maps marked * have been surveyed only in part.

SP 27	NW	Balsall Common	R.A.O. 1978–1980
	NE	Tile Hill	C.H.C. 1914, R.A.O. 1978
	SW	Chadwick End	R.A.O. 1978–1979
	SE	Kenilworth	R.A.O. 1977–1978
SP 25	NW	Tiddington	E.G.P. 1957, B.J.W. 1964, K.A. 1978, P.J.S. 1979
	*NE	Wellesbourne Hastings	K.A. 1978
SP 26	NW	Shrewley	K.A. 1978–1980
	NE	Warwick North	K.A. 1978
	SW	Norton Lindsey	J.B. 1977, P.J.S. 1979
	SE	Warwick South	J.B. 1977
*SP 35	NW	Morton Morrell	R.A.O. 1978
	NE	Harbury	E.A.E. 1958, J.B., R.A.O. 1978
SP 36	NW	Leamington	K.A. 1976–1977
	NE	Hunningham	K.A. 1976
	SW	Whitnash	J.B. 1976
	SE	Ufton	K.A., J.B. 1976
SP 37	NW	Coventry	R.A.O. 1978, 1986
	NE	Binley	M.G.S. 1977–1978
	SW	Stoneleigh	R.A.O. 1977
	SE	Ryton-on-Dunsmore	M.G.S. 1977–1978
SP 45	NW	Ladbroke	E.A.E., B.J.W. 1957–1958, J.B. 1978
	NE	Priors Marston	E.A.E., B.J.W., V.W. 1957–1958, J.B. 1977
SP 46	NW	Marton	K.A. 1976
	NE	Grandborough	K.A. 1977
	SW	Southam	J.B. 1976
	SE	Napton on the Hill	J.B. 1976–1977
SP 47	NW	Wolston	M.G.S. 1977, 1979
	NE	Long Lawford	M.G.S. 1977, 1979
	SW	Stretton-on-Dunsmore	M.G.S. 1977
	SE	Dunchurch	M.G.S. 1976
SP 55	NW	Hellidon	V.W. 1955, J.B. 1977
SP 56	NW	Braunston	R.H.H. 1950, K.A. 1977–1978
	SW	Flecknoe	V.W. 1955, J.B. 1977
SP 57	NW	Rugby North	R.H.H. 1950–1955, E.G.P. 1963, M.G.S. 1976, 1979
	SW	Rugby South	R.H.H. 1950, E.G.P. 1963, M.G.S. 1976

LIST OF OPENFILE REPORTS

Available from BGS Keyworth

Geological notes and local details for 1:10 000 sheets:

SP 26 NW, NE, SW, SE and parts of SP 25 NW and NE (Warwick and Hatton). K. Ambrose and P. J. Strange. 1982

SP 36 NW (Royal Leamington Spa). K. Ambrose. 1986

SP 36 NE (Offchurch). K. Ambrose. 1986

SP 37 NE (South-east Coventry). M. G. Sumbler. 1985

SP 37 SE (Bubbenhall). M. G. Sumbler. 1985

SP 47 NW, NE, SW, SE (Rugby West). M. G. Sumbler. 1985

SP 57 NW (North-east Rugby). M. G. Sumbler. 1983

SP 57 SW (South-east Rugby). M. G. Sumbler. 1982

CHAPTER 1

Introduction

The district around Warwick (Sheet 184) lies mainly in Warwickshire but includes parts of the counties of West Midlands and Northamptonshire. Outside the major centres of population, Warwick, Royal Leamington Spa, Kenilworth, Rugby and the southern portion of Coventry, most of the district is given over to agriculture.

Cement manufacture is a major industry at Rugby and Southam, but works at Harbury are now closed. Sand and gravel are worked at Brandon, Ryton-on-Dunsmore and Bubbenhall and were formerly quarried extensively between Wolston and Baginton and at Hillmorton. Workings from Binley Colliery, Coventry, extended beneath the district to Ryton-on-Dunsmore, but the pit was closed in 1963. However, exploration by the National Coal Board has proved workable coal resources over a wide area south of Coventry, and their development is currently being planned. Brick manufacture at Kenilworth and Napton on the Hill has ceased.

PHYSIOGRAPHY

The dominant SW – NE grain of the landscape reflects the south-easterly dip of the Mesozoic strata and the distribution of glacial drift. The highest ground lies in the south-east, where the Middle Lias escarpment rises to 200 m above OD near Hellidon and 205 m above OD at Beacon Hill, Upper Shuckburgh. Elsewhere only at Napton Hill and Bush Hill does the ground rise above 150 m above OD. Most of the country of the Leam and Avon valleys lies between 50 and 100 m above OD. North-west-facing scarps are formed by the Penarth Group between Chesterton Green and Birdingbury, and by the Blue Lias. The low-relief outcrop of the Mercia Mudstone and Lower Lias rises to more varied topography associated with glacial drift. The Dunsmore Gravel forms a plateau between Rugby and Princethorpe. Landscapes developed on the Enville Group have a predominant east – west grain, with the sandstones producing north-facing scarps and long dip slopes, and valleys cut in the mudstones.

GEOLOGICAL HISTORY

Although nothing is known of the geological history of the district prior to Upper Cambrian times, comparison with the area immediately to the north suggests that a volcanic Precambrian 'basement' is overlain by an almost complete Cambrian sequence (Eastwood and others, 1923). Upper Cambrian, together with Tremadoc rocks, proved only in boreholes, comprise dark grey mudstones and siltstones containing graptolites, trilobites and brachiopods, and were probably deposited in a shelf sea; the *Dictyonema flabelliforme* and *Clonograptus tenellus* zones of the Tremadoc have been iden-

tified. The structure of the Cambrian strata is imperfectly known. There appear to be several gentle folds, although with local dips as steep as 60°, produced by Caledonian earth movements. Following the orogeny the district became part of an uplifted east – west ridge known as the Mercian Highlands, and a long period of erosion ensued. Sedimentation did not resume until Westphalian times.

The Lower and Middle Coal Measures (Westphalian A, B and lower C) were deposited in an extensive coal swamp/deltaic environment. Epeirogenic movements resulted in cyclic sedimentation and in the splitting of individual coal seams. Grey mudstone and siltstone are the predominant lithologies. In the most stable areas the coal swamps were sufficiently persistent to allow the amalgamation of a number of coal seams, giving rise to the Thick Coal. During Lower and Middle Coal Measures times there was a slow extension of the area covered by coal swamps. The Aegiranum Marine Band is widespread, and is the only marine horizon so far identified in the Westphalian south of Binley, though others may exist. Shortly after this marine incursion, however, a regression of the swamps led to the deposition of the red mudstones and thin sandstones of the Etruria Marl Formation.

The overlying Halesowen Formation, the base of which is arbitrarily equated with the Westphalian C–D boundary, rests unconformably on the Etruria Marl, and marks the return of a coal-forming environment, although grey, fluviatile sandstone is the predominant lithology. These rocks overlap older Westphalian strata southwards, and rest directly on Cambrian shales. Several persistent coal seams in the south can be correlated with those of the Oxfordshire Coalfield. The Keele Formation comprises mainly continental, barren, red mudstones and sandstones with a few thin *Spirorbis* limestones. The overlying Enville Group consists of fluviatile, red-brown mudstones, siltstones and sandstones with subordinate breccias and conglomerates; the oldest rocks exposed in the district (Figure 1) belong to the upper part of the Coventry Sandstone Formation.

A jaw-bone of *Ophiacodon* from the Tile Hill Mudstone Formation has been assigned a late-Stephanian to early-Autunian age and is the only fossil of stratigraphical significance from the lower part of the Enville Group and the Keele Formation. The position of the Westphalian/Stephanian boundary is, therefore, uncertain. The upper part of the Enville Group (Kenilworth Sandstone and Ashow formations) has yielded a sparse, Lower Permian (Autunian) vertebrate fauna, and the base of the Permian is arbitrarily placed at the base of the Kenilworth Sandstone. No major unconformity has been detected within the Enville Group.

The Hercynian orogeny produced gentle folds on north – south axes in the Westphalian strata. Uplift followed by erosion laid bare Cambrian rocks before Triassic deposition began, and the consequent Triassic basal unconformity is less pronounced in the west than in the east. The Broms-

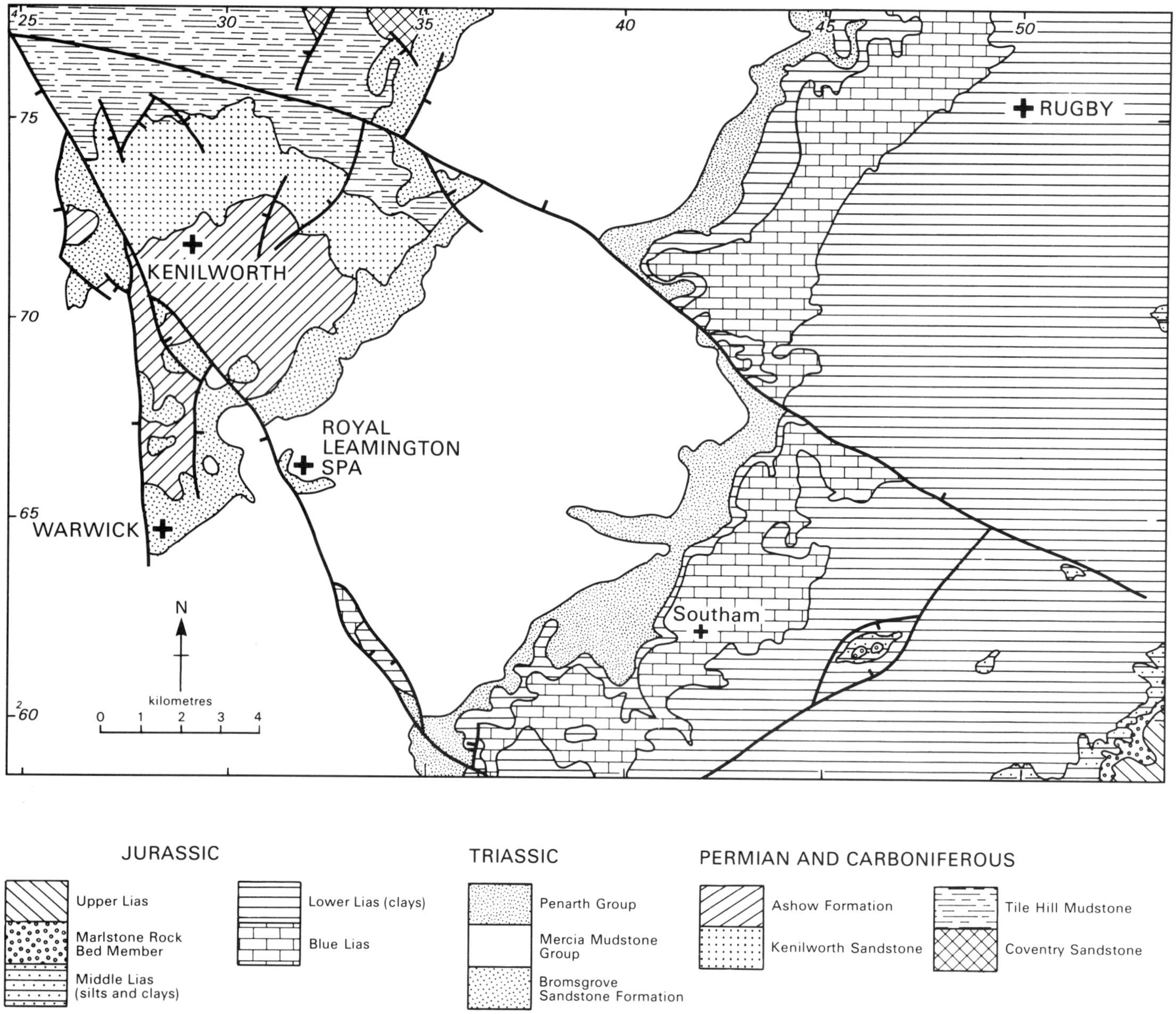

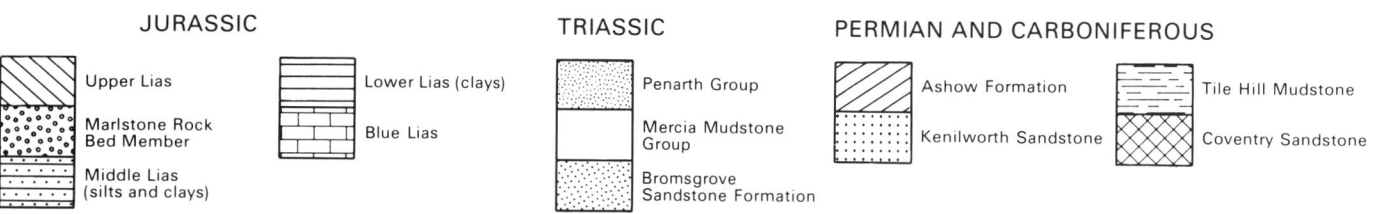

Figure 1 Geological sketch map of the Warwick district

grove Sandstone Formation comprises fluvial, fining-upwards sequences of brown sandstone and red mudstone, its false bedding indicating that the bulk of the sediment had a southerly or westerly derivation. West of the Warwick Fault, a sharp increase in thickness of the formation suggests that sedimentation was fault controlled. The predominantly red and argillaceous Mercia Mudstone was laid down in a continental basin where periodic floods deposited subordinate siltstones and sandstones. Less oxidising conditions were marked by deposition of the grey-green mudstones of the Blue Anchor Formation. Late in the Triassic a widespread marine incursion gave rise to the black mudstones of the Westbury Formation. The overlying pale grey-green calcareous mudstones of the Cotham Member are indicative of brackish water with a few marine interludes, and the limestones of the Langport Member mark the invasion of the

district by the Mesozoic sea.

Jurassic sediments accumulated in this sea, the shelly clays and limestones of the Lower Lias in deeper water than the succeeding silts, sandstones and ferruginous limestones. A return to deeper water is marked by the Upper Lias mudstones, but the overlying Northampton Sand indicates a temporary change to deltaic conditions.

The known glacial deposits of the district are apparently largely the products of a single glaciation (Shotton, 1953). The boulder clays contain erratics that suggest an easterly derivation for much of the ice. Sands and gravels were laid down by glacial meltwaters, and stoneless clays were deposited in glacial lakes. Late Pleistocene erosion has destroyed most of the glacial landforms, and glacial debris has contributed to widespread river terrace and alluvial deposits.

CHAPTER 2

Cambrian

Wherever proved, the pre-Carboniferous rocks beneath the district are of Cambrian age (here taken to include the Tremadoc Series); they are assigned to the Merevale Shales (Tremadoc Series). The Upper Old Red Sandstone in the Nuneaton Inlier north of Coventry is overstepped southwards by Carboniferous strata some 5 km north-west of Nuneaton (Taylor and Rushton, 1971, p.43), though in the Banbury district to the south Devonian strata lie between Cambrian (Tremadoc) and Carboniferous. In the present district, however, there is no record of this Devonian basement.

Cambrian rocks have been penetrated by several boreholes (Figure 2), but only to a maximum thickness of 33 m and no detailed correlation is possible. All known provings are in the Tremadoc Series, and the boreholes concerned are shown in Figure 2. The faunal zones proved show that the strata young into a NE–SW belt, presumed to be a syncline, although in the south-east of the district data are sparse.

Taylor and Rushton (1971) divided the Upper Cambrian part of the Stockingford Shales of the Nuneaton Inlier as follows:

Merevale Shales	90 m	Tremadoc Series
Monks Park Shales	80 m	
Moor Wood Flags and Shales	15 m	Merioneth Series
Outwoods Shales	300 m	

The gradational passage from the Monks Park Shales to the Merevale Shales lacks a distinctive fauna, so the boundary between the Merioneth and Tremadoc series cannot be identified precisely. Above these passage beds the Tremadoc rocks of the Nuneaton Inlier are poorly known, and a more useful comparison can be made with the divisions of the Shineton Shales of Shropshire (Stubblefield and Bulman, 1927), which are as follows:

Arenaceous Beds	
Shumardia pusilla Zone	
Brachiopod Beds	
Clonograptus tenellus Zone	Tremadoc Series
Transition Beds	
Dictyonema flabelliforme Zone (disconformity at base)	

The Tremadoc of the Warwick district includes the lowest three of the above divisions and continues the sequence downwards, as follows:

Clonograptus tenellus Zone:	with *Clonograptus* and *Adelograptus* species
Transition Beds:	*Clonograptus* is interbedded with *Dictyonema*, including *D. flabelliforme anglicum*
Dictyonema flabelliforme Zone:	*D. flabelliforme flabelliforme* and subspp. but no *Clonograptus*

'Basal Tremadoc':	lacking distinctive biostratigraphical characters

Probably a broad subcrop of Shineton Shales, up to 2000 m or more of Tremadoc strata, extends beneath central England (Stubblefield and Bulman, 1927, p.115; Cave, 1977, p.10; Bulman and Rushton, 1973).

The Tremadoc rocks beneath the Warwick district consist of mudstones with thin sandstone beds. The mudstones are grey, greenish grey, rarely dark grey, and in places are reddened immediately below the sub-Carboniferous unconformity. They range from fissile to massive and silty, and some contain micaceous and pyritic layers. The sandstones are generally thin (2 to 100 mm), fine grained and cross laminated. Some massive silty beds show cone-in-cone structure. Besides current structures such as ripple-drift, toolmarks and evidence of scour, the sediments show signs of penecontemporaneous soft-sediment deformation, such as small-scale load casts, graptolites deformed by mud-flow, and incipient fracture-cleavage passing laterally, within the span of the core, into polished bedding-surfaces (cf. Davies and Cave, 1976). Bioturbation is widespread and faint burrows, commonly pyritic, occur in the mudstones. The sandstones and siltstones show horizontal sinuous burrows (some resembling *Planolites*), vertical burrows (some in pairs), trails or tracks ('footprints' of arthropods?), scratch-marks, and pellets, some of which fill burrows (*Tomaculum*). Body-fossils are rare to common in the mudstones but were not observed in the sandy interbeds. Commonest are inarticulate brachiopods, especially small *Lingulella* and acrotretids. Sponge spicules and problematical dark traces of organic origin occur widely, but trilobites and other arthropods such as bradoriids are very rare. Graptolites are common and of the greatest biostratigraphical value.

The presumed 'Basal Tremadoc' beds were proved in six boreholes west of Coventry and comprise grey and dark grey mudstones with a few thin sandstones. Fossils include sponge spicules, small acrotretids and conodonts, especially '*Prooneotodus*' *tenuis*. North of the district the core of Berry Fields Farm Borehole [2499 8148] yielded numerous sponge bodies as well as spicules, acrotretids resembling small *Eurytreta sabrinae*, '*Prooneotodus*' *tenuis*, and the trilobite *Euloma sp.*, the latter suggesting a stratigraphical level above the Monks Park Shales; an analysis of the microflora by Dr R. E. Turner indicated an age no younger than lowest Tremadoc.

Strata of the *Dictyonema flabelliforme* Zone typically comprise pale grey silty mudstones with numerous sandstone interbeds, as in Crewe Farm, Stareton, Black Spinney, Southam, In Meadow Gate and Holmes House boreholes (see Figure 2). The *Dictyonema* are mainly those referred by Bulman (1954) to *D. flabelliforme flabelliforme*, and are generally associated with *Eurytreta sabrinae* and small *Lingulellae*, possibly *L. nicholsoni*. *Broeggeria*, *Linnarssonia* cf. *belti* and

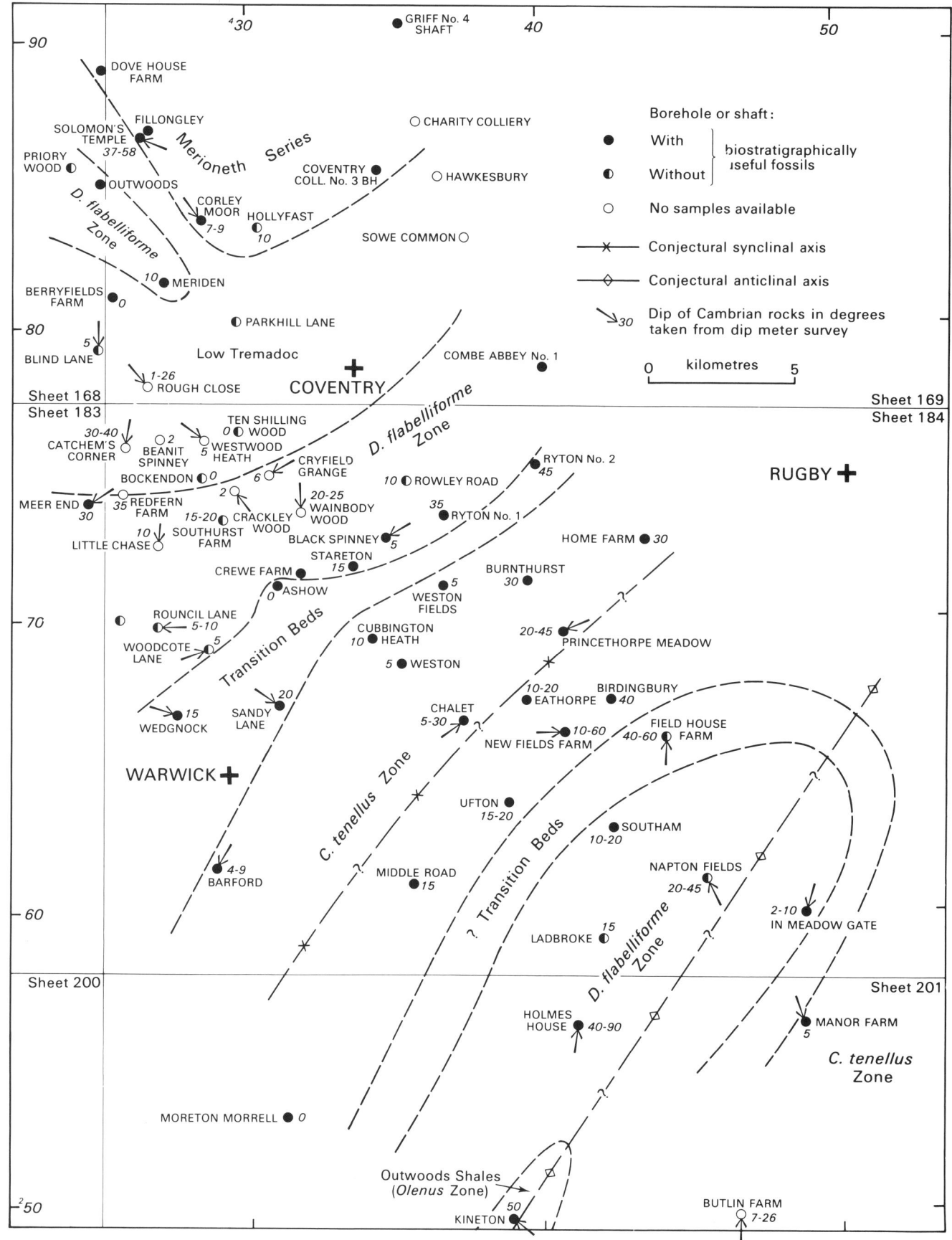

Figure 2 Cambrian faunal zones proved in shafts and boreholes

sponge spicules are rare, and no 'Prooneotodus' tenuis was observed. North of the district, Meriden Borehole [2682 8186] yielded *D. flabelliforme* aff. *sociale* (Bulman and Rushton, 1973, p.13), and Outwoods Borehole [2463 8528] numerous *D. flabelliforme belgicum*, some with short bithecae. The locations of these two boreholes close to subcrops of the Monks Park Shales point to a low horizon in the Tremadoc; hence the view that short bithecae in *D. flabelliforme belgicum* are an indicator of high Tremadoc rocks (Bulman and Rushton, 1973, p.8) is not supported, though the structure is too imperfectly known for it to be positively disproved. *D. flabelliforme patulum* occurred in Meer End and Combe Abbey No. 1 boreholes. Bulman and Rushton (1973, p.11) suggested a high Tremadoc horizon at the latter site, but a reconsideration of the associated shumardiid trilobites indicates strata below the *C. tenellus* Zone.

Grey mudstones with thin sandstones of the Transition Beds from the Brandon (Ryton No. 2) and Wedgnock boreholes showed the characteristic alternations between *Clonograptus tenellus* and *D. flabelliforme anglicum*. The lithologies and the associated fauna of small *Lingulellae* and rarer *Eurytreta sabrinae* most resemble those of the *D. flabelliforme* Zone. Unique to the Wedgnock core are conjoined valves of a *Septadella*. Sandy Lane Borehole yielded *D. flabelliforme anglicum* and numerous brachiopods similar to those at Wedgnock, but no *C. tenellus*, so reference to the Transition Beds is less certain. Ashow Borehole proved paler, greenish grey mudstones with fewer sandy interbeds, and with brachiopods similar to those of Wedgnock but including some *Linnarssonia* cf. *belti*; one anisograptid fragment and one specimen of *Shumardia curta* were noted. This core would probably have been referred tentatively to the *C. tenellus* Zone but for the fact that *S. curta* is known only from the *D. flabelliforme* Zone and the Transition Beds.

Strata of the *Clonograptus tenellus* Zone are mainly mudstones and silty mudstones, commonly greenish grey. Sandstones are fewer than in the *D. flabelliforme* Zone, and the rocks are more fossiliferous. Of the graptolites, *C. tenellus* is the commonest, and its subspecies *tenellus*, *sarmentosus* and *hians* have all been found. *Adelograptus* fragments are present but poorly preserved; *A. hunnebergensis* occurs at Ufton, and south of the district at Moreton Morell. *Lingulella* is common, being represented by small specimens, possibly of *L. nicholsoni*, together with a broad form which appears to be specifically different (Owens and others, 1982, p.22) and which seems particularly characteristic of the *C. tenellus* Zone. Acrotretids are common, especially *Eurytreta sabrinae* and *Linnarssonia* cf. *belti*, but *E. bisecta* and *Torynelasma sp.* occur rarely. *Broeggeria salteri*, apparently thicker-shelled than those in the Monks Park Shales, occurs sporadically, and *Palaeobolus quadratus* rarely. Sponge spicules are widespread but 'Prooneotodus' is rare, observed only at Weston Fields and Manor Farm. Trilobites are very rare, the only determinable example being a large but fragmentary *Niobella* from the Chalet core. A curious feature of the *C. tenellus* Zone is the presence of shapeless masses of dark, apparently organic, matter, flattened along the bedding and commonly associated with chitinozoa; these chitinozoa are *Lagenochitina esthonica* (about 1 mm long) and occur singly or in chains of several individuals. These associations have not been observed at lower horizons, but occur in all the *C. tenellus* Zone cores except that from Manor Farm. They are therefore taken as a local index of the *C. tenellus* Zone, and for this reason strata in the Birdingbury core are referred to that zone despite the absence of *Clonograptus*.

No beds higher than the *C. tenellus* Zone have been proved in the district. Rushton (1981) described *Kiaerograptus? quasimodo* from the Napton Fields core, and because no definite *Kiaerograptus* has been found in beds as old as the *C. tenellus* Zone, he tentatively suggested that the Napton Fields core represented a level in the middle or upper part of the Tremadoc Series. Although *K.? quasimodo* is now thought to be correctly referred to *Kiaerograptus*, the presence of a horizon above the *C. tenellus* Zone at Napton Fields appears anomalous because boreholes to the north-west (Southam), south-west (Holmes House) and east (In Meadow Gate) all proved the *D. flabelliforme* Zone; *K. quasimodo* may therefore be an early species of the genus.

CHAPTER 3

Igneous rock

In Ryton No. 3 (1948) Borehole [3631 7531], 9 cm of altered olivine-dolerite occurred beneath the presumed base of the Coal Measures at 934.4 m. The rock is medium greenish grey, with some small dark altered phenocrysts. In thin section (E 22439) it is composed of a mesh of unorientated, completely altered feldspar laths, with phenocrystic pseudomorphs after olivine. All the feldspars and primary ferromagnesian minerals are replaced by leucoxene, clay minerals, chlorite, carbonate and pyrite.

The rock was recovered from beneath a conglomerate containing pebbles of Cambrian rocks. Core recovery was poor and no unconformity was recognised above the olivine-dolerite, but the latter probably predates the Coal Measures. Mr R. K. Harrison reports that there are no mineralogical or textural similarities with the post-Cambrian lamprophyres or any of the Precambrian igneous rocks around Nuneaton.

CHAPTER 4

Upper Carboniferous (Westphalian) and ?Lower Permian (Autunian)

No Viséan or Namurian strata are known within this district, and the Westphalian beds rest directly on the Cambrian basement. Exploration to the north in the exposed Warwickshire Coalfield has been extended southwards and Westphalian A to Westphalian D strata are now well known beneath later cover (National Coal Board, 1985). This information supplements the older exploratory boreholes and workings from Binley Colliery (Figure 3). The Westphalian of the Coventry district has been described by Eastwood and others (1923) and Mitchell (1942).

A generalised sequence for the Westphalian and ?Autunian of this district is shown in Figure 4. The sequence consists mainly of mudstones, siltstones, sandstones and seatearths. The lower part and the Halesowen Formation are grey and coal-bearing; the Etruria Marl Formation is variegated red and green; the Keele Formation and Enville Group consist almost entirely of red measures, predominantly argillaceous in the former and predominantly arenaceous in the latter.

At the southern margin of the district the Halesowen Formation cuts down unconformably through Westphalian C, and eventually causes Westphalian D to rest on the Cambrian basement. This unconformity, while recognisable to the north, becomes more marked in its effect to the southeast of Leamington Spa.

WESTPHALIAN (COAL MEASURES)

The classification of the Warwickshire succession is shown in the accompanying table (Table 1). The exact chronostratigraphic horizon of the base of the Westphalian succession in the south of the Warwickshire Coalfield is uncertain, but is probably within the Lenisulcata or Communis chronozones. The Westphalian A/B boundary is taken immediately above the Thin Coal, the position where the Vanderbeckei Marine Band occurs in the northern part of the district and farther north. The B/C boundary is the Aegiranum Marine Band.

Table 1 Classification of the Westphalian of the Warwickshire Coalfield (in part based on Ramsbottom and others, 1978, pls.1–3)

Stages	Non-Marine Bivalve Zones (Trueman and Weir, 1946)	Marine bands	Warwickshire succession	Lithostratigraphic divisions	
				Mitchell (1942)	Stubblefield and Trotter (1957)
D	Prolifera		Enville Group (part) Keele Formation	Upper Coal	Upper Coal
	Tenuis		Halesowen Formation	Measures	Measures
C	Phillipsii		Etruria Marl Formation		
		Cambriense			
	Upper Similis-Pulchra				
B	Lower Similis-Pulchra	Aegiranum	Nuneaton Marine Band		Middle Coal Measures
				Productive Coal	
	Modiolaris	Vanderbeckei	Seven Feet Marine Band	Measures	
A	Communis		Unconformity		Lower Coal Measures
	Lenisulcata				
		Subcrenatum			

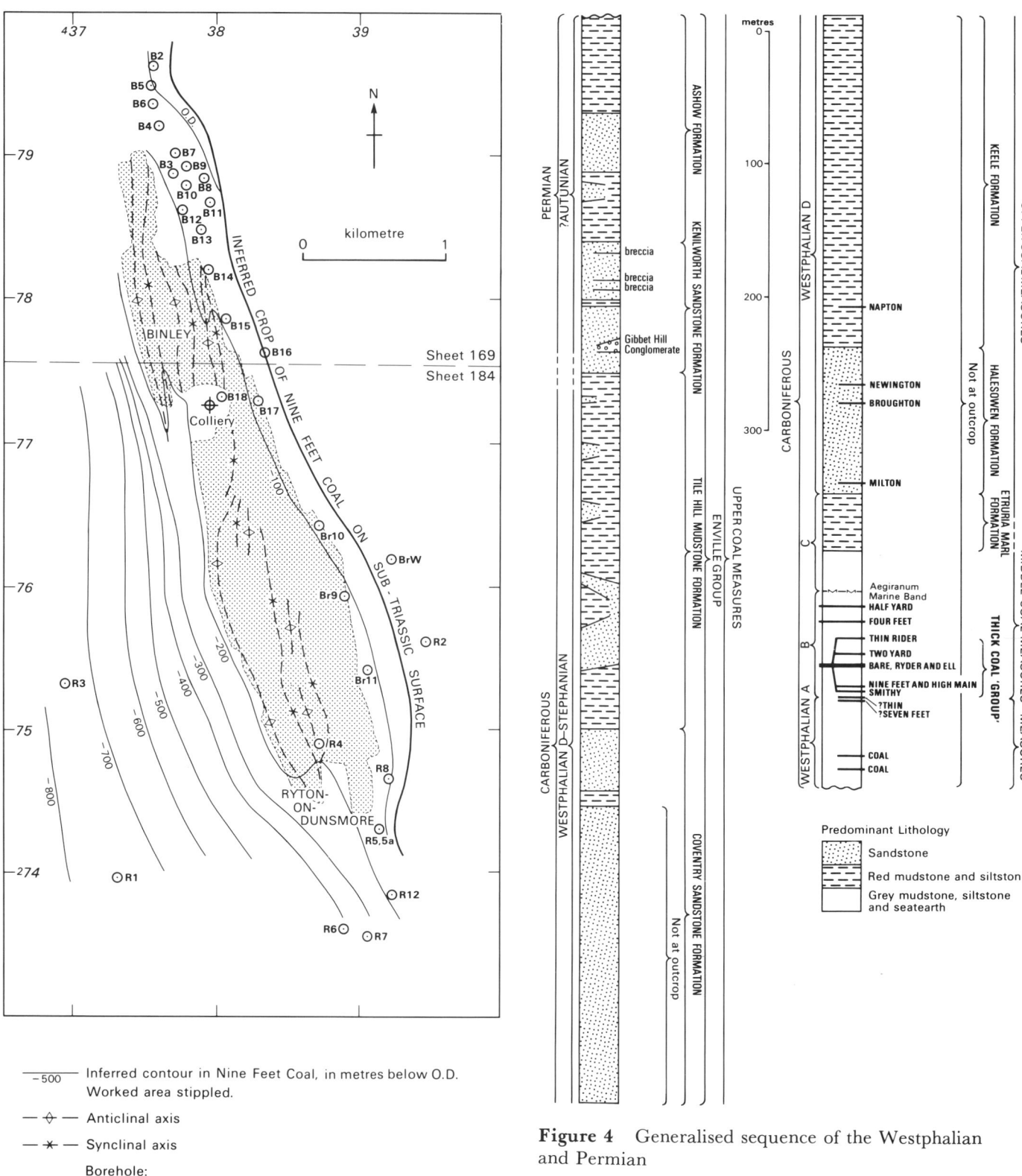

Figure 3 Binley Colliery: structure contours on Nine Feet Coal

Figure 4 Generalised sequence of the Westphalian and Permian

Due to the paucity of both marine and non-marine faunas, a precise chronostratigraphic classification of the upper part of the Westphalian of Warwickshire is impractical at present. The horizon of the base of the Etruria Marl Formation is variable, and it is convenient to take the base of the Halesowen Formation as the base of the Westphalian D stage, which is marked by an unconformity.

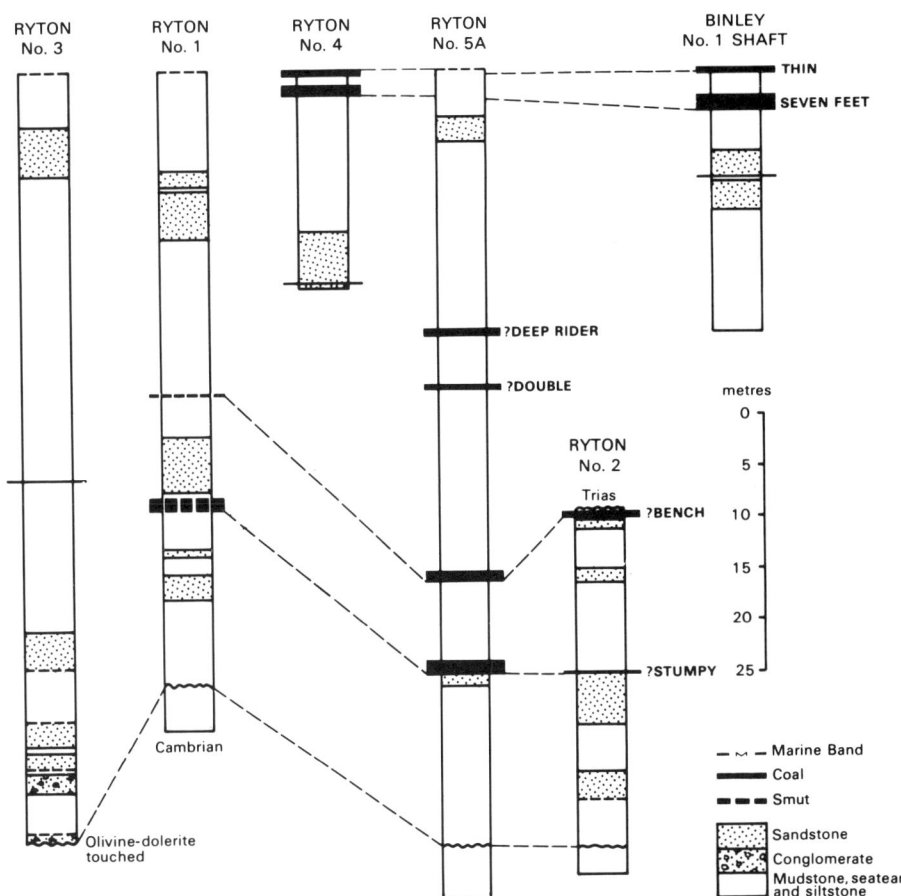

Figure 5 The Westphalian A succession near Binley

WESTPHALIAN A (Figure 5)

Until 1975 only three boreholes (Ryton Nos. 1, 3 & 5A) within the district had penetrated the full thickness of Westphalian A. Between 1975 and 1982 the National Coal Board drilled 45 more holes and an additional seven lie only just outside the confines of the district; all but one of these penetrated the whole of Westphalian A finishing in the Cambrian. In only five was the Vanderbeckei Marine Band identified, but in another eleven its position is marked by a thick ironstone band.

The thickness of Westphalian varies between 30 and 50 m, with local increases to 75 m and decreases to 25 m. The bulk of the strata are recorded as mudstones and seatclays; in some areas sandstones and conglomerate occur in the basal few metres. The unconformity with the underlying rocks is not always obvious since the top few centimetres of the Cambrian are commonly weathered and may contain Carboniferous rootlets.

In the northern part of the coalfield Westphalian A contains seams of coal—notably the Thin, Seven Feet and the Bench. The recent exploration has shown that these are thin and impersistent in this district, and positive seam identifications are not feasible due to the paucity of recorded nonmarine markers. The Princethorpe Meadows Borehole yielded *Carbonicola* cf. *pseudorobusta* and *C.browni* at 646.6 m, 40 m below the base of the Thick Coal Group, indicating mid-Westphalian A. The succession is characterised by abundant seatearths, many of which are brownish or putty-coloured, with abundant sphaero-siderite; for which deposition close to the landward margin of the coal-forming delta with frequent emergence is indicated. The Seven Feet and overlying Thin Seam are thickest in the north of the district where, in the shafts of the old Binley Colliery [3796 7726], the Seven Feet Seam is at 258.7 m depth and is split into a lower leaf of 0.66 m and an upper one of 0.48 m, separated by 0.03 m of seatearth: the Thin is taken to be a 0.46 m coal at 255.1 m depth.

WESTPHALIAN B (Figure 6)

Westphalian B includes the highest normally productive measures of the Warwickshire Coalfield, (but see Halesowen Formation). The thickness of the beds is about 55 m but increases to 70 m around Binley.

Measures below the Thick Coal Group

These measures consist of seatclays and mudstones, usually with dark mudstone towards the base. They are generally 4 to 5 m thick, but thin to only 1.8 m in Ryton No. 4 Borehole. The beds include, at their base, the horizon of the Vanderbeckei Marine Band which has been proved in the

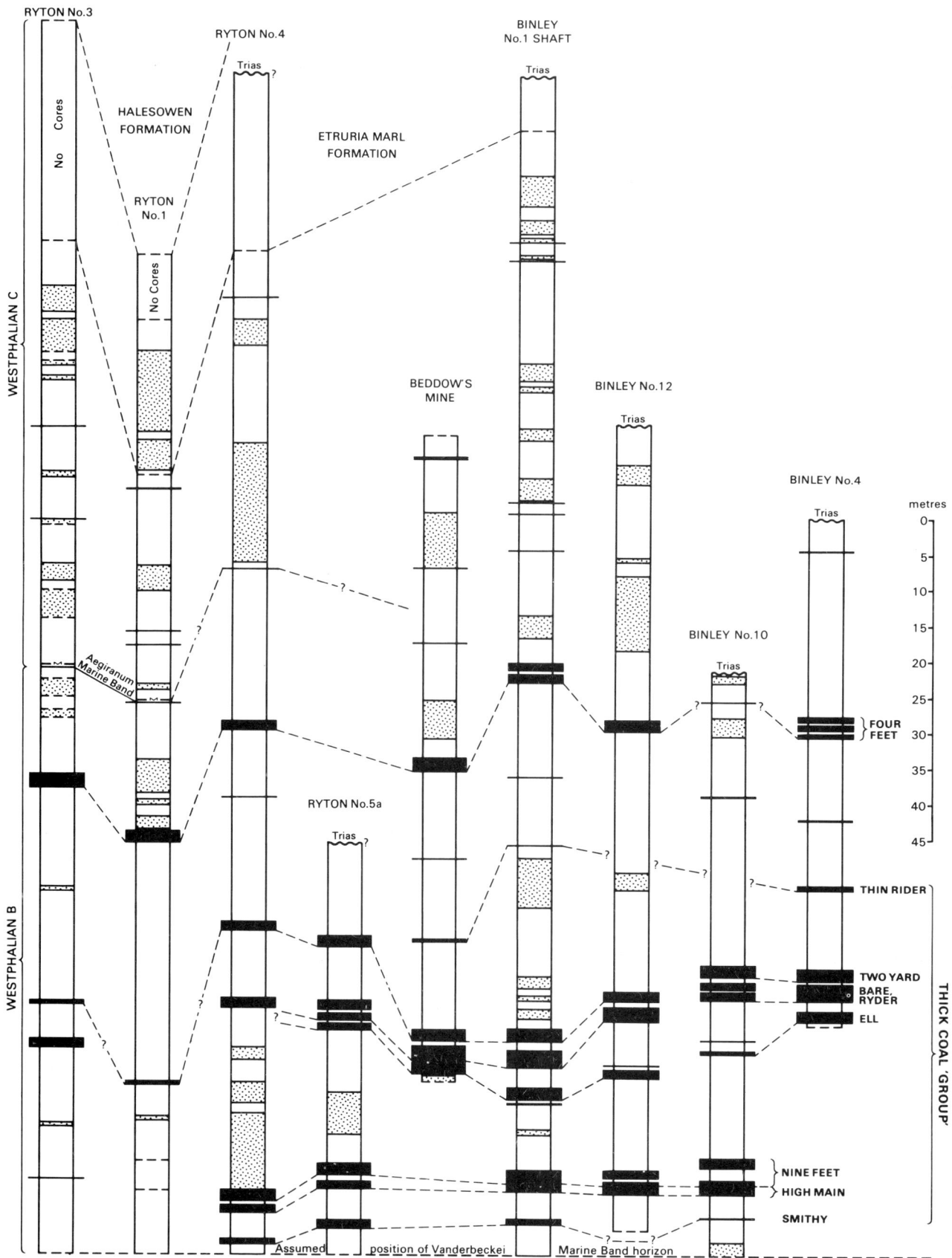

Figure 6 The Westphalian B and C successions near Binley. Borehole sites given on Figure 3: key given on Figure 5

region north of Coventry (Mitchell, 1942). The Marine Band was proved in five of the N.C.B. boreholes in the northern part of the district, where it had an impoverished fauna. In these boreholes, and in others in the Coventry district, it is accompanied by a distinctive ironstone band which is recognisable over a much wider area, perhaps extending as far south as Kenilworth.

Thick Coal Group

Over part of the Warwickshire Coalfield in the district to the north, a number of coal seams combine to form the Thick Coal (Cope and Jones, 1970), the constituent seams of which are at some point either in contact, or separated by no more than a thin dirt parting. Farther south the seams separate again, and in the Binley area, where they have been worked, they are spread out over a considerable vertical interval. The aggregate thickness of coal at Binley, is even greater than in the true Thick Coal region, reaching over 10 m in Binley Colliery Shafts. Mitchell (1942, p.8), considered that the Thick Coal comprised the Nine Feet (Slate), Ell, Ryder, Bare, Two Yard and Thin Rider Coals. Cope and Jones (1970, p.585) also included the Smithy (Lady) and High Main coals beneath the Nine Feet.

The recent exploration has shown that the splitting of the Thick Coal into leaves approximating to its constituent parts in the Warwick district is not simple (National Coal Board, 1985). For convenience a number of areas have been defined with distinctive seam sections; these are shown on Figure 7.

The correlation of the constituent leaves of the Thick Coal throughout the district is difficult, but has been achieved with certainty using chemical variation and coal macerals. The parts which have been identified are in descending order Two Yard, Bare, Ryder, Ell, Nine Feet (which splits into two), High Main and Smithy. The Thin Rider Seam which lies above the Two Yard Seam to the north is combined with the Two Yard Seam forming a widely recognisable top zone within that seam.

On the western margin of the coalfield, in part of the zone of Partial Thick Coal (see Figure 7), the lower part of the Nine Feet, the High Main and Smithy seams are not developed; the western side of the Prime Thick Coal zone indicates where all the seams are closely associated. South-eastwards there is, first a zone where many of the seams separate from each other but the dirt partings within the major leaves are no more than 30 cm thick, though those between the major leaves are much more, (Splitting Thick Coal); next a narrow zone of Split Thick Coal where there are large intervals between all the leaves; then an area (Recombining and Recombined Thick Coal) where the seams come together again. Beyond this to the southern margin of the district is a zone in which the seam deteriorates with failure of the middle leaves associated with offlap of the lower part and onlap of the upper. This variation is shown diagrammatically in Figure 7. Widely spaced boreholes to the south of Warwick and Leamington Spa show that the Thick Coal fails in a similar manner here, but the Two Yard persists as a thin seam for a substantial way to the south. Current mining technology limits the coalfield that is assessed as 'workable'.

Thick Coal to Aegiranum Marine Band

The roof of the Thin Rider usually consists of black shale with ironstone bands. A fauna collected in Grinley's Mine [c.378 772] yielded *Anthracosia atra*, *A.* aff. *atra*, *A.* cf. *acutella*, and *Anthracosphaerium propinquum*. In Beddows Mine [3715 7712], the following fauna was found above the Thin Rider: *A. atra*, *A. simulans* and *Naiadites sp.* In Ryton No. 3 Borehole, a similar fauna comprised *Anthraconaia cymbula*, *Anthracosia atra*, *A.* cf. *acutella* and *A. simulans*, together with fish remains including *Elonichthys* scales. Similar *Anthracosia atra* faunas occur in the roof of the combined Two Yard and Thin Rider seams farther south in the Princethorpe Meadows, Middle Road and Southam boreholes, the latter including *A. atra*, *A.* aff. *concinna*, *A.* aff. *fulva* and *Naiadites obliquus* from 882.0 to 883.0 m. These faunas indicate a late Westphalian B, Lower Similis-Pulchra Chronozone age.

The interval between the Thin Rider and the Four Feet coals averages about 23 m in thickness near Binley, thinning southwards to 5–15 m. Above the black shale roof of the Thin Rider, the measures consist mainly of seatclays and shaly mudstones. In the Binley shafts, and in borings to the north, a thin coal (0.15 to 0.33 m) has been proved consistently about 10 m above the Thin Rider.

The roof of the Four Feet commonly contains an *Anthracosia atra* fauna identical to that above the Thin Rider/Two Yard. The next coal upwards is the Half Yard, some 25 m above the Thick Coal. It occurs in a more arenaceous succession and has little or no associated fauna. The highest coal is thin and impersistent, lying 30–40 m above the Thick Coal. It marks the top of Westphalian B, and carries the Aegiranum Marine Band in its roof.

WESTPHALIAN C (Figure 6)

Aegiranum Marine Band to Etruria Marl

The base of the Westphalian C stage is marked by the Aegiranum Marine Band which appears to persist over the northern part of the district. To the south the succession becomes much more arenaceous and contains a number of discontinuities; accordingly it is difficult to be certain of the correlation between the widely spaced boreholes. It is probable that before the southern margin of the district is reached, Westphalian C is completely cut out by the unconformity at the base of Westphalian D. Because of this the thickness of the stage is very variable from a maximum of around 90 m at Ryton to zero near Southam.

The richest development of the Aegiranum Marine Band in the district occurred in Rouncil Lane Borehole where the following diverse fauna was recovered; *Lingula mytilloides*, *Orbiculoidea* cf. *cincta*, *Dictyoclostus sp.*, *Neochonetes granulifer*, *Plicochonetes waldschmidti*, *Tornquistia diminuta*, *Myalina sp.*, cf. *Pernopecten sp.*, *Phestia sharmani*, *Euphemites anthracinus*, *Retispira sp.*, *Hollinella* cf. *ulrichi*, *H. claycrossensis*, *Roundyella sp.*, *Ditomopyge sp.*, *Serpuloides stubblefieldi*, conulariid and fish debris. Such faunal associations are typical of nearshore environments in contrast to the more offshore goniatite dominated assemblages seen in some other regions at this horizon. In the Beddows Mine exploration drift [3731 7715] (Mitchell, 1942, p.11), the Aegiranum Marine Band yielded *Serpuloides?*, chonetoid fragment, *Lingula mytilloides*, *Orbiculoidea* cf. *nitida*, *Phestia acuta*, and conodonts including

Hindeodella sp. and *Ozarkodina sp.* In Ryton No. 3 Borehole, a fauna from the Aegiranum Marine Band from between 775.4 and 777.5 m, included *Hyperammina sp.*, *L. mytilloides*, *L. sp. nov.*, *Levipustula sp.* (fragmentary), *O. cf. nitida*, *Productus?* (fragmentary), *Dunbarella sp.*, *Pernopecten carboniferus*, *Schizodus antiquus*, *Coleolus sp.* and an *Elonichthys* scale. The Marine Band was also proved in Ryton No. 1 Borehole, but no fossils were retained.

Above the dark grey mudstone of the Marine Band, the measures consist mainly of grey mudstones and seatearths, with some thin coals (rarely up to 0.3 m) and sandstones. In Ryton No. 3 Borehole, *?Euestheria* occurred in brown and dark grey mudstone immediately above 0.03 m of coal about 34 m above the Aegiranum Marine Band.   MGS

Etruria Marl Formation

In the northern part of the Warwickshire coalfield the Etruria Marl consists of brown and red variegated 'marls' with lenticular beds of sandstone (Eastwood and others, 1923; Mitchell, 1942). In this district the formation is by comparison less distinctive. A sequence of mudstones, siltstones and sandstones, mainly greyish in colour, but always variegated with red and brown, occurs between the grey Productive Coal Measures beds and the overlying Halesowen Formation; this is taken to be the Etruria Marl. The sandstones are commonly coarse and gritty, but the conglomeratic espleys, which characterise the Etruria Marl farther north, are rare. The thickness of the formation varies from about 30 m in the north to nothing around Leamington. There also appears to be some transgression of the base of the Formation, and it seems probable that while it is visually similar to the Etruria Marl of the type succession it is not of the same age; possibly it is related to the unconformity at the base of Westphalian D.

WESTPHALIAN D

The precise ages of the formations above the Westphalian C are not firmly established but appear at least to be Westphalian D in part, though the upper beds may well be Stephanian. The Halesowen and Keele formations are here regarded as Westphalian.

Halesowen Formation

The Halesowen Formation is dominated by thick beds of grey, coarse, feldspathic sandstone, interbedded with grey mudstones, siltstones, seatearths and rare coals. Many of the boreholes penetrating the full thickness of the formation have cored only its lower part, and the top of the formation is im-

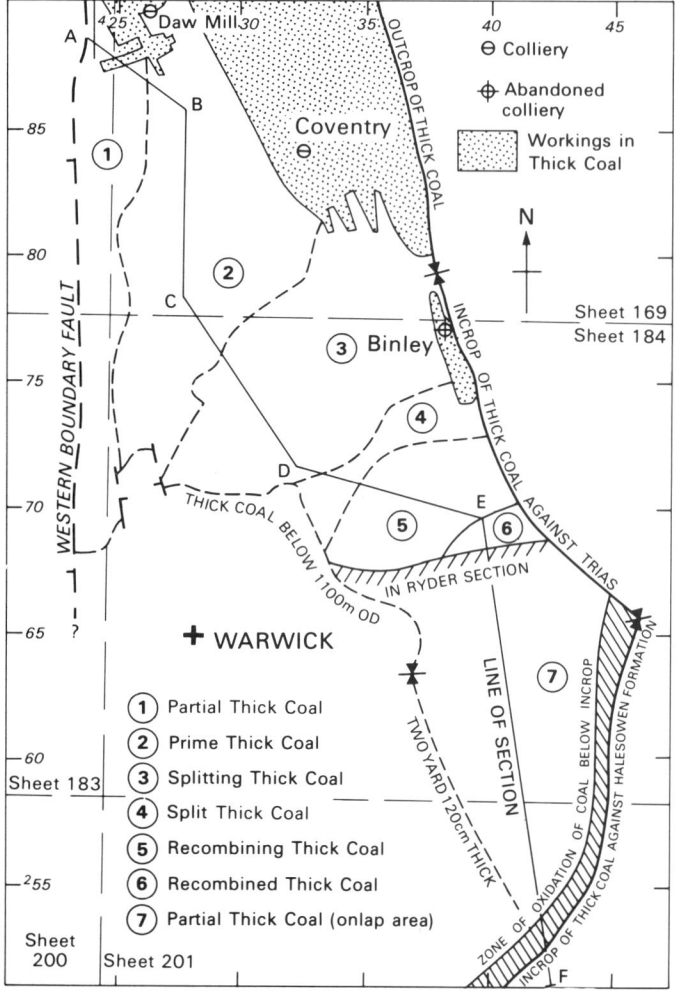

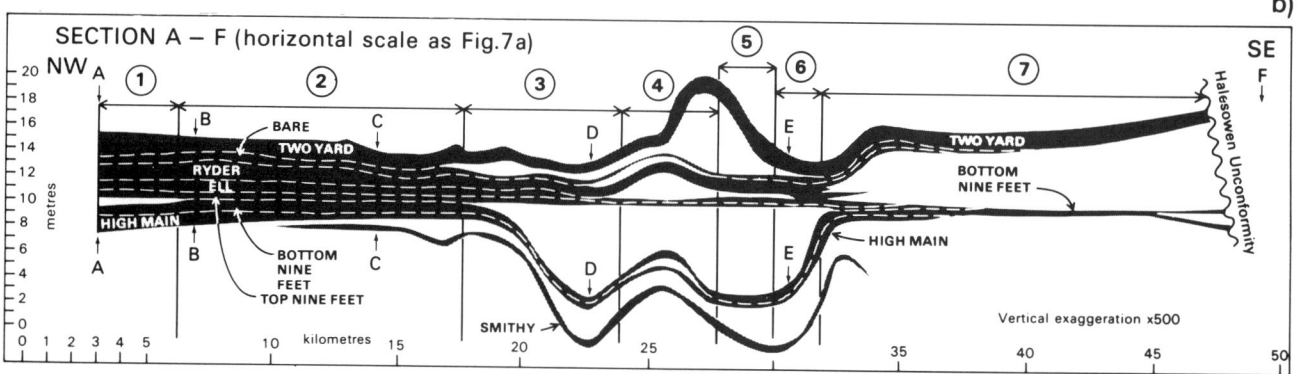

Figure 7a,b   Seam structure of Thick Coal (data from National Coal Board)

precisely known. In the north the formation is about 125 m in thickness, increasing southwards to about 200 m. The southwards thickening is not due to an increase in thickness of the lower beds, since the lowest seam of coal remains at a constant distance above the unconformity which marks the base of these beds. In the south of the district the formation rests directly upon Lower Palaeozoic rocks.

Recent exploration drilling by the National Coal Board has established the persistence of three coal seams in the south of the district, the Milton, Broughton and Newington seams in respective upwards order. The coals correlate respectively with the No. 9, Nos. 5 and 6 and No. 3 seams of Withycombe Farm Borehole [4319 4017] near Banbury (Poole, 1978, p.15) (Figure 8). *Leaia minima*, recovered 5 m below the Milton in the In Meadow Gate Borehole, and *Anthraconauta phillipsi*, *A. tenuis* and *Carbonita spp.* occurring between leaves of the Milton at 1172.0–1176.0 m in Sandy Lane Borehole, indicate a Westphalian D age.

Keele Formation

The Keele Formation has not been generally cored in this district, and is known mainly from chipping samples and geophysical borehole logs. In the district to the north it consists predominantly of red-brown mudstones with subordinate sandstones, and a few thin beds of Spirorbis limestones (Eastwood and others, 1923). A thin coal, the Napton Seam, occurs about 30 m above the base of the Formation in Napton Fields and In Meadow Gate boreholes, and may correlate with the No.2 seam of Withycombe Farm Borehole (Poole, 1978, p.14). From the few cores available there appears to be a gradual transition from the predominantly grey Halesowen Formation to the red measures of the Keele. The geophysical logs from the boreholes show a remarkably high gamma radiation over a narrow zone at about this level and this has been taken as the base of the Keele Formation although it may not correspond precisely with the lithological boundary: the geophysical correlation of the sandstone beds in the overlying beds seems to confirm that this is a widely persistent marker. The Formation is distinguished from the overlying Coventry Sandstone Formation by its more argillaceous nature and top of the Formation is thus rather indefinite. As thus defined the formation is about 230 m in thickness in the north, increasing to a maximum proved thickness of over 300 m around Birdingbury. *Anthraconaia pruvosti* and *Anthraconauta* aff. *phillipsi*, recovered 34 m above the Napton Seam in the In Meadow Gate Borehole, are of Westphalian D age. RAO, MGS

WESTPHALIAN D TO AUTUNIAN: ENVILLE GROUP

The Enville Group of the Warwickshire Coalfield was first defined as covering all the pre-Triassic strata that succeed the Keele Formation (Eastwood and others, 1923, p.77). This definition was modified by Shotton (1929, p.169) to include all the pre-Triassic rocks above the Tile Hill Beds of the earlier writers.

The outcrop of the Enville Group forms an inverted triangle with its apex at Warwick. It forms part of the southwards plunging syncline of the Warwickshire Coalfield,

and the lithological units have outcrops convex slightly northwards. The Group continues under Mesozoic cover beneath the western two-thirds of the district, but is probably absent northeast of the Princethorpe Fault and east of Willenhall.

The Geological Society's proposed schemes of correlation for the Permian (Smith and others, 1974) and the Silesian (Ramsbottom and others, 1978) have not clarified the position in Warwickshire. In the former, the Ashow, Kenilworth and Gibbet Hill subdivisions are considered to be groups (following Shotton, 1929, p.170): in the latter the Gibbet Hill and lower named subdivisions are thought of as members of an Enville Formation. The subdivisions used in this account are shown in Figure 4.

The Enville Group comprises red and red-brown mudstones, siltstones and sandstones with subordinate breccias and conglomerates. Four formations (the Coventry Sandstone, Tile Hill Mudstone, Kenilworth Sandstone and Ashow formations) have been recognised at outcrop and in geophysical borehole logs: they have a total thickness of about 900 m (Figure 8).

The base of the Permian has been placed by the Geological Society at the base of the 'Kenilworth Breccia Group' (Smith and others, 1974; Ramsbottom and others, 1978). This conclusion is founded on the supposed correlation of the Kenilworth and Clent Breccias (Shotton, 1929, p.201), and on the assumption that the unconformity at the base of the latter marks the base of the Permian. Due to the thin and impersistent nature of the breccias in the Kenilworth Sandstone the base of Shotton's 'Kenilworth Breccia Group' can be recognised only locally and does not provide a satisfactory position for the boundary between two systems. In contrast, the base of the Kenilworth Sandstone Formation (as defined below) can be traced over wide areas and is easily recognisable in downhole geophysical logs. It is consequently taken here as a convenient, if arbitrary, base to the Permian in the light of present day knowledge.

The occurrence in the Kenilworth Sandstone of the pelycosaurs *Sphenacodon brittanicus* and *Haptodus grandis* (Paton, 1974) and the amphibian *Dascyceps bucklandi* (Paton, 1975) lends support to the Permian age for the upper part of the Enville Group. All these species are confined to Autunian strata. A jaw-bone of the pelycosaur *Ophiacodon* discovered in the Coventry Sandstone '¾ mile north-west of Coventry' (Murchison and Strickland, 1840, p.347) is assigned a late-Stephanian to early-Autunian age by Paton (1974). Less well attested are the views of Haubold and Sergeant (1973, p.908) and Haubold and Katzung (1975, p.118), based on scanty evidence from reptile footprints, that the base of the Permian should be placed at the base of the Enville Group.

There is no chronologically significant fossil evidence as to the age of the Enville Group below the Coventry Sandstone, but the general concensus is that it is uppermost Carboniferous (Westphalian D or Stephanian). Dix (1935) reviewed the floral evidence and placed the whole of the Enville Group in the Permian, but no significant plant fossils occur below the Tile Hill Mudstones. Three specimens of 'Strophalosia' from the Enville Group figured by Howell (1859, p.32) have been reidentified as the gymnosperm seeds *Cardiocarpon reniforme* and *C. ottonis* (Cox, 1953).

Petrography of the Enville Group

The majority of Enville Group sandstones which have been examined in thin section are polylithic arenites. The grains are predominantly angular or subangular and moderately well sorted. Quartz and quartzite are the most important constituents, the latter, together with micaceous siltstone, sometimes attaining 40 per cent of the total (e.g. E 51562; Kenilworth Sandstone from Motslow Hill). Of the more unstable rock fragments, cloudy mudstones are the most frequent. Small amounts of heavily altered rhyolite, andesite and basalt are widespread, indicating a persistent igneous source. In contrast chlorite-mica-schist occurs only rarely (e.g. E 51561; Kenilworth Sandstone from Stoneleigh).

Mineral grains other than quartz are present in only minor or accessory quantities. They include microcline, albite, muscovite, biotite, amphibole, zircon and rare garnet.

Calcite is the usual cement in these rocks ranging from very large plates in which the grains almost float (E 51562) to a mosaic of very small plates (E 51561).

A somewhat different assemblage of rock fragments occurs in the breccia and associated sandstone (Tile Hill Mudstone) exposed in the railway cutting near Berkswell Station [2506 7740]. In the breccia (E 52685) altered igneous rocks are much more common than is usual in the Enville Group. Mr R. K. Harrison reports that these include felsite, trachyte, granophyre and possible andesite and rhyolite. Micritic and sparry limestone, a crinoid ossicle and chert were also observed. Among the mineral grains plagioclase and microcline were much more common than usual. In the sandstone (E 52686) chlorite-mica-schist was a common constituent. A similar assemblage of rocks was found in a conglomerate at Baginton (E 52683) [3388 7563] which probably lies at about the same horizon.

Details of the heavy mineral contents of Enville Group rocks have been published by Fleet (1925, 1927) and Richardson and Fleet (1926). Green reduction spots in the mudstones are described by Mykura and Hampton (1984).

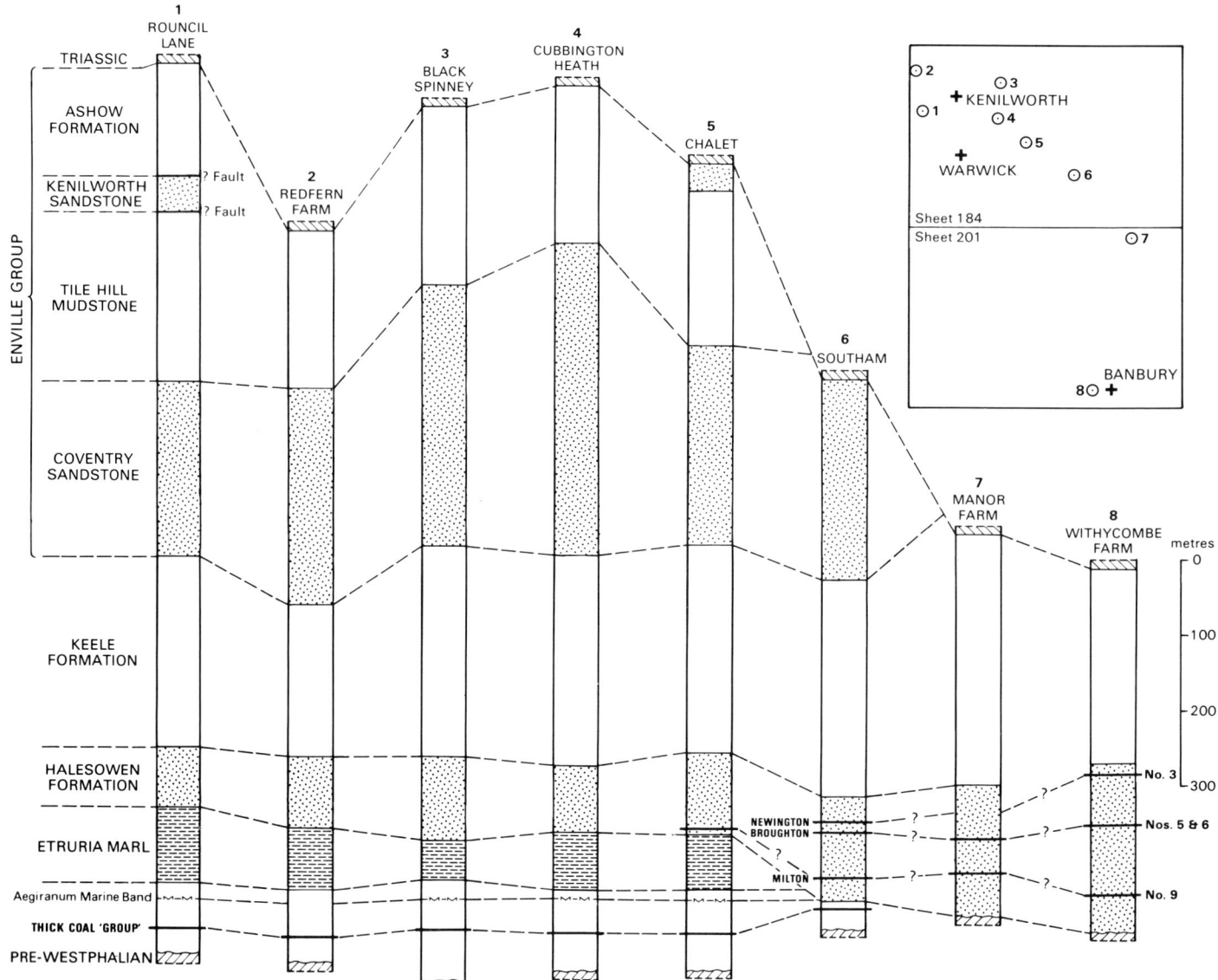

Figure 8   Comparative Westphalian sections in selected boreholes

Provenance

The provenance of the pebbles in the Enville Group is discussed by Shotton (1929, pp.190–198) who concludes that the majority were derived from the Precambrian and Lower Palaeozoic rocks of the Lickey Ridge, (the only exception being those in the Corley Conglomerate). The assumption that the Lower Carboniferous pebbles, especially limestone, have a westerly source cannot be proved. These pebbles may equally probably have been derived from the northern end of the Warwickshire Coalfield.
equally probably have been derived from the northern end of the Warwickshire Coalfield.

Coventry Sandstone Formation

The Coventry Sandstone Formation spans the interval between the top of the Keele Formation and the base of the Tile Hill Mudstone Formation. It includes the Arley/Exhall Conglomerate at the base, the Corley Conglomerate and, at the top, the Allesley Conglomerate (Eastwood and others, 1923). The Arley/Exhall and Allesley Conglomerates pass laterally into sandstone, and the Coventry Sandstone is best defined as a predominantly arenaceous formation lying between two formations composed predominantly of mudstone; this definition can be used in classifying geophysical logs of the exploratory boreholes.

The formation is about 350 m thick in the district to the north (Sheet 169). In the present district geophysical logs suggest it is about 300 m thick.

Details

Only the highest 100 m of the Coventry Sandstone crops out in this district, and the lower part, including the Corley Conglomerate, is proved only in boreholes. Most exposures are of typical Enville red-brown sandstone. Eastwood's 6-inch map of the area around Whitley Common records an extensive outcrop of 'Corley Conglomerate'. There are, however, good reasons to doubt this correlation. Almost no conglomerate was seen by Eastwood, and he inferred the outcrop from the presence of pebbly soils. Excavations and boreholes at the Whitley Interchange [347 772] have since shown up to 4.5 m of glacial gravel overlying sandstone with thin beds of conglomerate. Moreover, Eastwood's map shows only sandstone in the long railway cutting outside Coventry Station [338 780], but if this horizon is through the outcrop of the Corley Conglomerate, coarse conglomerate would be expected to feature prominently in the exposure particularly since Shotton (1927) found that the Corley Conglomerate was coarsening eastwards. He also found that the most easterly outcrop of Corley Conglomerate was composed almost entirely of Upper Silurian sandstone pebbles while the only present exposure of conglomerate in this area, at Whitley Bridge [3487 7721], contains a variety of other pebbles including limestone, quartzite and red sandstone, and apparently lies near the top of the Coventry Sandstone, probably at the level of conglomerates in the Mount Nod boreholes (Shotton, 1933). If this view is correct, then the outcrop of the Corley Conglomerate, which certainly passes through central Coventry, must be offset north-eastwards by a fault. Such a fault, with an easterly downthrow has been traced between Green Lane and Top Green, Coventry, and its conjectural extension northwards would cut the Corley Conglomerate beneath the centre of Coventry.   RAO

Ryton No. 3 Borehole [3694 7531] proved about 260 m of Coventry Sandstone to a depth of 317.4 m. The record of chipping samples shows mainly 'marl and sandstone', but includes 15 m of conglomerate at 90.6 m which is almost certainly the Corley Conglomerate. A small amount of conglomerate proved in dominantly

sandstone sequences in old borings at Willenhall Bridge [3584 7670] and Whitley Well [3577 7673] may also be the Corley Conglomerate (Richardson, 1928, pp.182–184). Maidwell (1910) stated that the beds dipped west at about 15° in headings driven from the Whitley Well.

In the Lodge Farm water borehole at Baginton [3640 7410], 64.8 m of Coventry Sandstone were cored below the base of the Tile Hill Mudstone at 89.9 m. Thick beds of sandstone, mainly red but some grey, make up 70 per cent of the sequence, and thinner beds of red mudstone the remainder.   MGS

Tile Hill Mudstone Formation

The stratigraphical limits of this formation have been given by Shotton (1929, pp.171–172). The Tile Hill Mudstone comprises the predominantly argillaceous sequence between the highest mappable persistent sandstone of the Coventry Sandstone and the predominantly arenaceous Kenilworth Sandstone Formation. It includes some impersistent sandstones, especially towards the middle of the sequence, and a thin conglomerate occurs at Beechwood. The thickness of the formation is about 280 m, close to Shotton's (1929) estimate of 274 m. Thicknesses obtained from geophysical borehole logs vary from 250 to 300 m in the north of the district to as little as 150 m around Warwick and Leamington Spa (Figure 8).

Good exposures of the mudstones are rare. The mudstones are red-brown, well bedded, locally silty and with green reduction spots. Sandstones are more significant than suggested by Shotton (1929). Several crop out in the south western part of Coventry, though all die out rapidly along the strike. Some are of typical 'Enville' lithology: red, coarse, massive and fairly soft. Others, more characteristic of this formation, are hard, flaggy, red-brown or green, thin, fine-grained, calcareous, and interbedded with mudstone. There are numerous exposures of such sandstone, interbedded with mudstone and siltstone, in a stream between Wolfe Road, Canley [2900 7775] and Canley Ford [3127 7700].

Details

Eastwood and others (1923, p.87) and Shotton (1929, p.172) allude to an impersistent sandstone with layers of conglomerate in a quarry at Beechwood [2691 7713]. The quarry has since been backfilled, but the sandstone makes a pronounced north–south feature from Little Beanit Farm [256 766] to Hodgett's Lane [259 773]. The sandstone is not exposed in a railway cutting to the north, due to an intervening east–west fault, which moves the outcrop 900 m to the west where the following section was measured in the railway cutting [2508 7740]:

	Thickness m
Sandstone; brown, flaggy, false-bedded, with scattered clasts of mudstone and red chert; 2–3 mm pebbles of grey quartzite at base; channelled into underlying bed	1.5
Silty sandstone; red-brown, flaggy, with impersistent grey-green partings; cut out to east by bed above	0–1.1
Sandstone; brown, in beds up to 1.5 m thick with lenses of breccio-conglomerate up to 0.2 m; the latter contain abundant mudstone clasts up to 3 cm in diameter and small clasts and pebbles of white quartzite and grey sandstone; pebbles and clasts of quartzite and chert are scattered throughout the sandstone	seen 3.9

Under the Kenilworth Sandstone, Kenilworth Pumping Station No. 4 Well [2956 7283] recorded 'Sandstone and conglomerate' in two beds 11 and 12 m thick at depths of 215 and 228 m respectively. Both here and at Beechwood the conglomerates lie about 150 m from the top of the Tile Hill Mudstone. A few small pebbles were also encountered in a sandstone at a depth of 154 m at Green Lane Pumping Station [3220 7598] (Butler, 1946, p.40), but this level is estimated to be about 250 m below the top of the formation.   RAO

The basal beds of the Tile Hill Mudstone Formation were proved in Lodge Farm Borehole. They comprised 19.5 m of 'red marl', including 4.7 m of interbedded red sandstone and mudstone. South of Bubbenhall Bridge, four boreholes proved Tile Hill Mudstone beneath First Terrace gravels; the deepest bore [3503 7249] proved 'hard red and green marl' to 6.1 m, below the base of the First Terrace deposits at 1.5 m. Debris from trenches south of the bridge [3523 7234; 3535 7526; 3524 7230] all indicated the presence of mudstones beneath the terrace deposits. The river bluff [3510 7295] west of Bubbenhall Bridge is largely made up of tough red mudstones and siltstones, with thin sandstone beds. A roadside exposure [3522 7301], 80 m NE of the bridge was in 2 m of red, medium- to coarse-grained, gritty sandstone; dips varied from 15° to the south, to 10° to the north.

A borehole [4199 6396] at Long Itchington Cement Works records dark marl beneath the Bromsgrove Sandstone from 175.3 m to the bottom of the hole at 182.9 m; these beds probably lie near the base of the Tile Hill Mudstone.   MGS

Kenilworth Sandstone Formation

This formation, approximately 100 m thick, includes the 'Kenilworth Breccia Group' and the 'Gibbet Hill Group' as defined by Shotton (1929). The base of the former can be recognised only locally due to the impersistence of the breccia bands.

The base of the Kenilworth Sandstone is marked by the incoming of thick, massive sandstones forming a strong north-facing scarp from Hurst Farm though Gibbet Hill to Stoneleigh. West of Hurst Farm the scarp is broken by faulting, but the base of Kenilworth Sandstone can be traced as far as the Warwick Fault. The top of the formation is taken at the onset of the thick mudstones of the Ashow Formation.

Red, massive, and commonly soft sandstones form the bulk of the formation with subordinate thin lensing mudstones. Towards the base the Gibbet Hill Conglomerate is locally present, and there are thin lenses of breccia mainly towards the top. Shotton (1929) identified only two breccia bands, respectively the 'Upper' and 'Lower' breccias, and this terminology has been retained in the details below for ease of cross reference to his paper. There are few localities where any appreciable thickness of sandstone is exposed. In 1978, however, a pipe-trench was excavated across virtually the whole width of the Kenilworth Sandstone outcrop between Gibbet Hill [307 748] and Kingswood Farm [314 728]. This confirmed that most of the sandstones are rather soft, at least to depths of 2–3 m. A few very hard, flaggy, calcareous, lustre-mottled bands were encountered. Red-brown mudstones with green spots were also encountered, and varied from blocky to thinly bedded.

Details

GIBBET HILL CONGLOMERATE   The Gibbet Hill Conglomerate was first described by Shotton (1929, pp.173–175). He estimated that the conglomerate occurred approximately 60 m above the base of the 'Gibbet Hill Group', but this is an over-estimate and the interval is only about 25 m. The outcrop of the conglomerate is terminated by a fault [2940 7458] between Cryfield Grange and Crackley Wood, and was not located farther west. Sandstone with a few small rounded and angular pebbles occurs in a cutting on the A452 at Drakes Hill [2690 7396], Shotton (1929, p.175, loc.a) ascribed this horizon to the Gibbet Hill Conglomerate, and claimed that the angularity of the pebbles indicated closer proximity to a westerly source than those at Gibbet Hill. However, the conglomerates at Gibbet Hill are thicker and have larger pebbles than the pebbly sandstone at Drakes Hill, and the latter is probably a local facies, lying only a few metres above the base of the Kenilworth Sandstone.

At the type locality on Gibbet Hill the sections measured by Shotton (1929, p.174) are now considerably obscured. In the 'Middle Quarry' [3045 7521] the following section was measured:

	Thickness m
Conglomerate; not clearly seen	-
Sandstone; red-brown to brown, coarse, pebbly, false-bedded especially towards base	0.6
Conglomerate; mainly well rounded pebbles of low sphericity, 6 to 7 cm maximum diameter; larger pebbles are usually Enville sandstones or mudstones	seen 2.0
(Shotton records a further 1.4 m of conglomerate)	

Shotton identified the pebbles; there is a predominance of 'Valentian' sandstones or quartzites and red Carboniferous cherts; other quartzites, vein quartz, Carboniferous limestones and Precambrian tuffs occur in minor amounts.

In 1986 a 16 m-long roadside excavation immediately east of this quarry showed a 1 m bed of mudstone into which the lower conglomerate of the quarry was channelled. The base of the mudstone was not exposed, but it probably occurs as a lens within the conglomerate. A 3 m-long lens of sandstone occurred within the mudstone.

The outcrop of the conglomerate is easily traced southwards from Gibbet Hill, passing just east of Cryfield Garage Farm: up to 2 m are well exposed in a stream [2986 7451] SSW of the farm. No trace of pebbles or conglomerate could be found east of Gibbet Hill Farm and the conglomerate has apparently almost lensed out. However, loose blocks of conglomerate were seen in the railway cutting at Gibbet Hill, and conglomerate brash appears in a field at Wainbody Wood Farm [3145 7435]. This latter occurrence was considered to be 'Kenilworth Breccia' by Shotton (1929, p.177, loc.d), but the conglomerate contains almost entirely rounded pebbles, unlike typical 'Kenilworth Breccia', and it lies close to the base of the Kenilworth Sandstone, well below the level of the breccia at Kenilworth.

The outcrop of the conglomerate could not be traced east of a fault at Wainbody Wood Farm, but the bed is said to have been found in excavations at New Era Farm [3173 7414]. Two conglomerate beds 0.3 and 0.6 m thick were encountered at depths of 4.6 and 5.5 m respectively in a borehole at Kingswood Farm [3200 7325].

Weston Colony Borehole [3662 6907] proved 19.7 m of strata, assigned here to the Kenilworth Sandstone, below the Bromsgrove Sandstone. They consisted of red sandstones and mudstones with two layers of conglomerate; the thickest is 2.13 m and may be the Gibbet Hill Conglomerate.

BRECCIAS   Most of the breccias are restricted to the immediate neighbourhood of Kenilworth. To the east they are rare, and the Kenilworth Sandstone is, in practice, indivisible.

At least two breccias occur at levels below those described by Shotton. The lower of these was encountered within a sandstone exposed in a trench west of Westley Bridge [3136 7367]. The following downwards sequence was measured: conglomerate with flattened pebbles of mudstone, siltstone and ironstone and rare quartzite and red chert 0.2 m; very hard calcareous sandstone with partings of mudstone clasts and a few pebbles 0.2 m; breccia with quartzite and chert pebbles and clasts 0.1 m. This band is about 30 m above the base of the formation, and so is very near to the horizon of the Gibbet Hill Conglomerate. A breccia recorded by Shotton (1929, p.177, loc.c) in 'the bank of a pool 1000 yards (305 m) south of Gibbet Hill cross-roads' may be at, or a little above, this horizon, and is well below the level of Shotton's 'lower breccia' at Kenilworth.

The second breccia is about 65 m above the base of the Kenilworth Sandstone. It was recorded by Richardson and Fleet (1926, p.289) from the Kenilworth Pumping Station No. 2 Bore [2956 7283] at a depth of 12 m. The same breccia occurs in Holly Cottage Well [2831 7321] where 4.3 m of sandstone at 13.4 m includes bands of pebbles and clasts.

Lower Breccia   Shotton (1929) mapped two breccias at Kenilworth; the lower, forming the base of his 'Kenilworth Breccia Group', and a second said by him to be 33 m above, but more probably nearer 25 m, although poor exposure and uncertain correlation between outcrops makes thickness estimates unreliable. The Lower Breccia lies about 70 m above the base of the formation. Shotton (1929, p.177, loc.a) described 0.5 m of 'coarse breccia, weathered, in a sandy matrix' in Finham Brook, north of Kenilworth Castle [2777 7259]; this locality is now obscured. Richardson and Fleet (1926, p.177) and Shotton (1929, p.177, loc.b) noted breccia in Clinton Lane, Kenilworth [2788 7265]; this locality is also now obscured, but breccia fragments were dug in house foundations of the west side of Clinton Lane [2785 7265].

The best present exposures are at Love Lane Quarry [2866 7284], originally ascribed by Shotton to the Upper Breccia. The following section is still accessible (cf. Richardson and Fleet, 1926, p.289; Shotton, 1929, p.179):

	Thickness m
Sandstone; massive, red-brown, false-bedded, with scattered pebbles	4.0
Breccia; with pebbles and clasts of soft red mudstone, hard grey sandstone and some limestone and chert, rapidly lensing to a pebbly bedding-plane	0–0.45
Sandstone; as above	4.0
Breccia; as above but more persistent; passes into muddy sandstone locally	0.4

The 'Small Love Lane Quarry' (locality d of Shotton), is now obscured.

The 'Lower Breccia' has not been found east of Love Lane, and the eastward continuation of its outcrop shown by Shotton is based on miscorrelation.

Upper Breccia   The only clear exposure of this breccia is at Castle Quarry, Kenilworth [2778 7191] (Plate 2) where the following sequence was measured (cf. Shotton, 1929, p.117):

	Thickness m
Sandstone; massive, brown with scattered small pebbles; uneven base, passing down to	0.8
Breccia; coarse with clasts up to 55 mm diameter	seen 1.2

In the west of the quarry the sandstone includes more pebbles, and passes laterally into coarse breccia at the top while the breccia recorded above passes into 1.5 m interbanded sandstone and brec-

cia. The beds immediately underlying the breccia are seen at the western end of the quarry where the face comprises 3 m of flaggy brown sandstone with rare pebbles.

The breccia was formerly exposed in small quarries [2830 7260; 2837 7260] north-west of St Nicholas Church, Kenilworth, (Richardson and Fleet, 1926, p.289), its outcrop having been displaced northwards by a fault. It is not known farther to the east of this locality.

Beds in the upper part of the Kenilworth Sandstone, including the probable horizons of the Lower and Upper breccias, form a prominent scarp at Stoneleigh and are exposed in several small quarries. In one on Motslow Hill [3294 7233] 2 m of massive and false-bedded brown sandstone overlie 2 m of hard, grey, coarse, massive sandstone with a thin band of abundant mudstone clasts. Other small exposures in the neighbourhood of Stoneleigh confirm that mudstone clasts are common in these sandstones.   RAO

Ashow Formation

The Ashow Formation is equivalent to the Ashow Group of Shotton (1929), and comprises all the Enville Group above the Kenilworth Sandstone. It is predominantly argillaceous, but contains several thick sandstones. The total thickness of the formation is about 170 m. The upper part of the formation around Warwick is poorly exposed, and occurs in fault-bounded inliers. As a result a detailed succession and thickness remains somewhat uncertain.

Details

The base of the formation is marked by the incoming of a sequence of mudstones 50 to 65 m thick, divided in places by a sandstone up to 15 m thick. These are the 'Whitemoor Marls' and 'Whitemoor Sandstone' named by Richardson and Fleet (1926, pp.297–298) after the former Whitemoor Brickworks, Kenilworth [295 717]. Up to 10 m of mudstones towards the base of the formation are exposed at the former Cherry Orchard Brickworks, Kenilworth [295 722]. They are red-brown, blocky, with abundant green reduction spots, and interbedded with siltstones. A few lenses of sandstone, up to 0.5 m thick, occur; they are usually red-brown but some are pale grey-green. Similar beds slightly higher in the sequence were formerly exposed to a depth of 8 m in the adjacent Whitemoor Brickworks. Sun cracks, rain-drop impressions, ripple marks, reptilian footprints and Walchia piniformis have all been recorded there (Richardson and Fleet, 1926, pp.297–298; Shotton, 1929, pp.179–181]. Only about 4 m of mudstone are still visible and pass laterally into sandstone. This latter has an upper portion, 1.7 m thick which is brown, flaggy, and fine grained, with undulating bedding and load casts. The lower portion, of which only 2 m are exposed, is massive and red-brown. In the south the sandstone becomes flaggy with many green partings and abundant load casts picked out in green sandstone (Plate 3). Mudstones overlie the sandstone, and were dug to a depth of 8 m in a drain at Victoria Spinney [306 721]. The outcrop of these mudstones was crossed by a trench [310 717] near Crewe Farm, which proved the presence of thin intercalations of sandstone.

The major sandstone above these mudstones is about 60 m thick, and forms the long dip-slope on which Ashow is built. An almost complete traverse of the outcrop was provided by a trench west of Ashow. The sandstone is soft, flaggy, cross-laminated and deeply weathered, with a few hard calcareous bands.   RAO

The following section of part of this sandstone was exposed [311 693] ESE of Chesford Bridge. All the sandstones were medium or fine grained, and all the strata red-brown except where stated.

Plate 2   Breccia in Kenilworth Sandstone; Castle Quarry, Kenilworth.   A13100

	Thickness m
Sandstone, flaggy; mudstone clasts increasing towards base	1.9
Sandstone, more massive than bed above; scattered calcareous nodular masses	1.0
Sandstone; pinkish brown massive; calcareous masses as in bed above; erosive base (cutting out to west the 6 beds below within 30 m horizontally	1.5
Sandstone; flaggy	0–0.8
Sandstone; massive, with calcareous nodules	0–0.8
Sandstone cross-bedded	0–0.8
Sandstone; massive	0–1.1
Siltstone	0–0.4
Mudstone; with sand lenses	0–0.2
Mudstone; with sand lenses	0.4
Sandstone; orange-brown, with red mudstone lenses	0.9

Beds higher in the Ashow Formation are rarely exposed. About 5 m of brick red mudstone outcrop in a steep bank [290 675] on the west side of the A46 Warwick bypass. The top 0.5 m immediately below the Bromsgrove Sandstone, was smooth and blocky. Below this, the mudstones are silty, micaceous and fissile with fine laminations, irregular green staining and reduction spots.

Beds still higher in the sequence were exposed in an old pit [2883 6674] west of Guys Cliffe Chapel. The section is:

	Thickness m
Sandstone; purple-brown in top 0.1 m, becoming grey-green below, medium- to fine-grained, flaggy weathering	0.2
Mudstone; dark red-brown, silty, finely laminated with pale sandy laminae	0.7
Sandstone; red-brown, fine-grained, silty, micaceous, flaggy; ripple drift and micro-cross-bedding dipping NE	0.25
Mudstone; red-brown, blocky, silty	0.1

Boreholes and temporary exposures in the Woodloes housing estate [28 66] at Warwick indicate a dominantly mudstone sequence with thin interbedded siltstones and sandstones. Several boreholes, mainly in Warwick and Royal Leamington Spa have penetrated the Ashow Formation (Appendix 1): the southernmost is Tachbrook Mallory Boring [3226 6225]. All show sequences similar to those described above.   K A

Plate 3   Load casts in sandstone; Ashow Formation, Whitemoor Brickworks, Kenilworth.   A13093

CHAPTER 5

Triassic

The basal division of the Triassic rocks of the district is the Sherwood Sandstone Group, represented only by the Bromsgrove Sandstone Formation (formerly Lower Keuper Sandstone or Keuper Sandstone). Within the overlying Mercia Mudstone Group (formerly Keuper Marl) the only named subdivisions recognised at outcrop are the Arden Sandstone and the Blue Anchor Formation (formerly Tea Green Marl): in the Home Farm Borehole [4317 7309], however, several of the formations defined in the Nottingham area by Elliott (1961) have been distinguished (Table 2). The Penarth Group is divisible into the Westbury Formation below and the Lilstock Formation above; it includes beds formerly known as the Rhaetic. The Lilstock Formation comprises a lower Cotham Member and an upper Langport Member (formerly White Lias). Over much of the district it was not possible to separate the Westbury Formation from the Cotham Member owing to landslip, and the two have been shown together on the 1:50 000 map.

The age of the lower two groups is imprecisely known, but that of the Penarth Group is better documented. The terminology and likely stage correlations (based on Warrington and others, 1980) are shown in Table 2.

SHERWOOD SANDSTONE GROUP

Bromsgrove Sandstone Formation

The Bromsgrove Sandstone crops out in irregular faulted areas between Warwick and Burton Green, as a continuous, almost drift-free strip extending north-east from Warwick to Bubbenhall, and in faulted ground much mantled by drift between Bubbenhall and Pinley. Its base constitutes a major unconformity. The formation mainly comprises cross-bedded sandstones with subordinate mudstones. The former are mostly buff or pale grey-green, generally well sorted, medium- to fine-grained and micaceous; they contain many erosion and lateral accretion surfaces. Intraformational conglomerates (with clasts of mudstone and siltstone) and mudstone-pellet beds are common in the lower part of the formation. The mudstones are red-brown, and form mappable units west of Kenilworth and near Baginton, Bubbenhall, Whitley and Cubbington. A thin, purple-red, sandy mudstone lies at the base of the formation around Blackdown.

Over most of the district the Bromsgrove Sandstone ranges from 25 to 35 m in thickness; the minimum known is about 20 m in the south-east. West of the Warwick Fault, Hatton Mental Hospital No. 2 Borehole [2498 6795] proved a thickness of 68 m (Butler, 1946). Westward thickening continues and in the type area at Bromsgove the formation is at least 368 m thick. The sharply increased thickness west of the Warwick Fault suggests fault control. The Bromsgrove Sandstone of Royal Leamington Spa is the source of the well known spa waters (Richardson, 1928, pp.121–133).

Table 2 Correlation of lithostratigraphy and chronostratigraphy for the Triassic of the Warwick district

Rhaetian	Lilstock Formation	Langport Member Cotham Member
	Westbury Formation	
Norian	Blue Anchor Formation	
	Glen Parva Formation	
	Trent Formation	
Carnian	Edwalton Formation	Arden Sandstone, Cotgrave Skerry
Ladinian	Unnamed red mudstone	
Anisian	Bromsgrove Sandstone Formation	
Scythian		

The Bromsgrove Sandstone of Warwick is famous for its vertebrate remains, discovered in many of the quarries worked during the 19th century. Coton End Quarry has yielded many teeth and vertebrae together with rare fragments of skulls and other bones. The relative profusion of specimens from a normally unfossiliferous formation attracted many workers during the last century. Paton (1974) and Walker (1969) have reidentified many of the specimens and given full references to earlier work. They concluded that both the amphibian and the reptile fauna indicated a position within the range of late Anisian to earliest Carnian, an early to mid-Ladinian age being the most probable. If this is so, the Bromsgove Sandstone of Warwick probably represents only the upper part of the formation, for in the type area at Bromsgrove (Warrington and others, 1980, table 6) the formation ranges down into the late Scythian. No palynomorphs have been recovered from the Bromsgrove Sandstone in the district.

Ten thin sections (E 51570–2, E 51574–6, E 52926, E 52946–8) of sandstones were examined by Mr G. E. Strong. Quartz is the dominant detrital constituent. All samples were feldspathic, some containing up to 50 per cent primary detrital K-feldspar. Grains range from angular to rounded; those of K-feldspar are usually well rounded and commonly with secondary overgrowths. The degree of sorting is variable, being poor in the conglomeratic sandstones. The minor and accessory detrital grains include much muscovite, with some biotite, zircon, pyrite, epidote, and goethite. Several slides show intergranular pore spaces. One exhibits lithification by grain contact welding; others are cemented by feldspar or sparry calcite. Analysis by SEM multichannel analyser showed the oxide coatings commonly present on detrital grains to contain major Mn and Fe, with Al, Si, K, Ca and minor Ba; X-ray powder diffraction analysis by Mr R. J. Merriman indicated mainly an amorphous manganese oxide. The coated grains are mostly in cross-bedded sandstones, and usually concentrated in layers parallel to bedding. Bands of coated grains show disruption by dewatering structures in an old quarry near Blackdown Hill [322 687]. The oxide may have come from diagenetic solution of ferromagnesian minerals (S. Burley, personal communication).

The formation exhibits cyclic sedimentation, a complete cycle comprising locally pebbly sandstone, with an erosive base, passing upwards into mudstone. Such complete cycles are most evident in the upper part of the formation, where they are up to 7–8 m thick. They are similar to the alluvial cycles described by Allen (1965). The major sandstones formed in migrating river channels. They are commonly cross-bedded, and some show slumping and distortion of foresets, dewatering structures, load casts and flame structures. Current directions obtained from dip azimuths of foresets (Figure 9) indicate that most sediment was derived from the south and west; this agrees with Warrington's (1970) observations. A few possible soil horizons are present. The mudstones formed largely as overbank material deposited during floods. Such conditions of deposition accord with the lenticular form of individual sandstone and mudstone units.

Details

Bubbenhall area to Pinley and the Home Farm Borehole

In a quarry [3452 7290] 300 m south of Chantry Heath Wood 1.5 m of pale brown, flaggy, micaceous sandstone overlie 1.7 m of pale brown to white, massive, micaceous sandstone with muddy partings; the upper unit shows false bedding dipping north-east, and contains many mudstone and siltstone clasts. RAO

A ditch [3527 7134] near Waverley Farm exposes 0.4 m of red and grey-green, mottled mudstone resting on sandstone. 'Fairies Hole' [3561 7228], a small man-made cave south-west of Bubbenhall Church, contains 2 m of massive, level-bedded, medium- to fine-grained, pink-brown sandstone, and a nearby old building-stone quarry [3586 7231] is in 5 m of massive, pale pink-brown sandstone whose cross-bedding indicates derivation from the south and west. Just west of Bubbenhall Bridge 0.4 m of pale buff, fine-grained, micaceous sandstone [3508 7297] lies close above the Enville Group; a borehole immediately to the south-east proved pale brown, silty sand resting directly on the Enville Group; in neither place was a basal conglomerate present.

Immediately south of the Princethorpe Fault, the disused and largely overgrown Rock Spinney building-stone quarries showed [3620 7357]:

	Thickness m
Sandstone; soft, buff to white, thinly bedded, fine-grained	0.8
Sandstone; massive, pink, with large load-casts or erosional scours on base	1.2
Sandy mudstone; red	0.5
Sandstone	seen

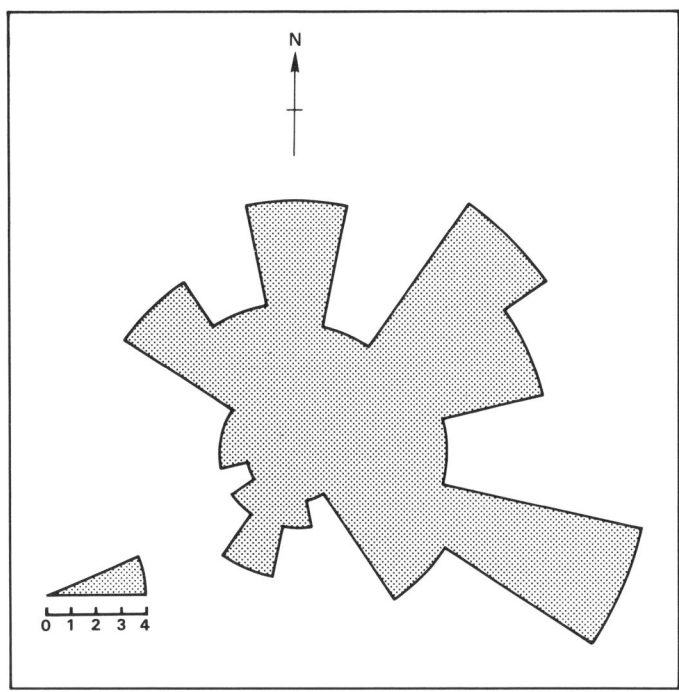

Figure 9 Dip azimuths of foresets in the Bromsgrove Sandstone Formation

A line of four temporary pits [3556 7697 to 3581 7694] south-east of Whitley Hospital all exposed sandstone with subordinate mudstone; the deepest [3556 7697] showed:

	Thickness m
Clay; red-brown, fissured (mudstone) with scattered grey-green mottles at depth; very sandy in basal 0.3 m with buff greenish grey streaks; passing to	1.0
Sandstone; soft and pink-brown (weathered) near top, hard and grey below, massive with a few thin beds of green-grey and red-brown mudstone; dip up to 2° to east or south-east	2.1

A marly conglomerate occurred at the base of the Bromsgrove Sandstone in headings driven from the Whitley Well [3577 7673] (Maidwell, 1910).

In the Home Farm Borehole [4317 7309] (Sumbler, 1980), about 40 per cent of the formation consisted of mudstone (Figure 10), with more in the lower part than in the upper. The sandstones were mostly grey to buff with a feldspar cement; the thicker ones had sharp erosive bases, and many contained layers of intra-formational, mudstone-flake conglomerate. Rare quartz pebbles were present, and also small nodules of reddish brown gypsum. The basal 0.72 m sandstone was conglomeratic, with small pebbles (mostly less than 1 cm) of white, grey and brown quartz, and large pebbles (up to 8 cm or more) of volcanics, metavolcanics, shales and subgreywackes. The mudstone units, commonly 1 to 2 m thick, were usually dark red-brown in colour, and some were silty or sandy. Many were micaceous and most were massive. Laminated layers, some disrupted by mudcracks, occurred in places, especially in the upper parts of the units, and some of the mudstones contained nodules of reddish brown gypsum. At 235.22 m the base of a thin sandstone cut through a gypsum nodule, indicating that the gypsum formed penecontemporaneously with deposition. MGS.

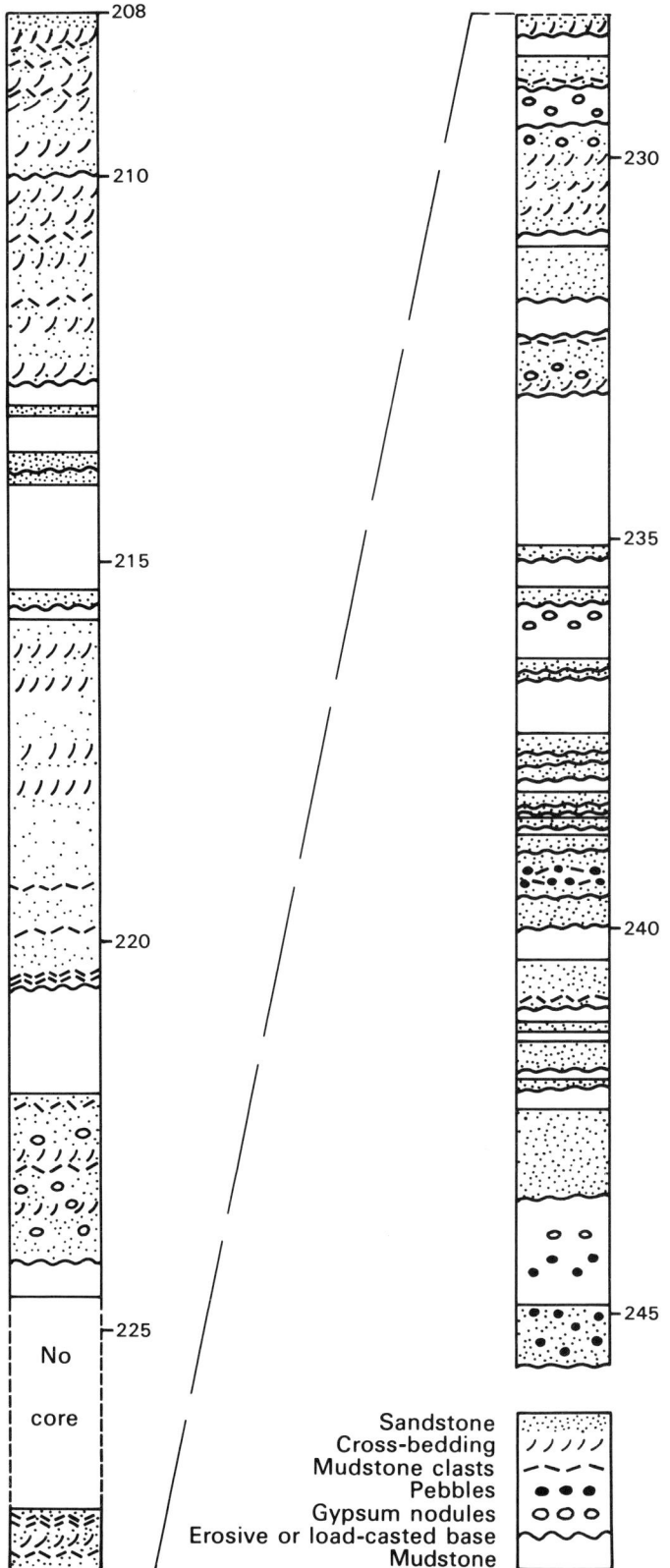

Sandstone
Cross-bedding
Mudstone clasts
Pebbles
Gypsum nodules
Erosive or load-casted base
Mudstone

Figure 10 The Bromsgrove Sandstone Formation in Home Farm Borehole

Warwick to Royal Leamington Spa

The best of the numerous exposures in this area are listed below, and the more important are described in detail (see also Figures 11 and 12).

	Locality	NGR	Thickness exposed m
1	Old Quarry, Blackdown Hill	3228 6880	5
2	A444 road, Blackdown Hill	317 687	3.4
3	Old Quarry, Blackdown Nursing Home	312 680	4
4	Cliff, south bank River Avon, Old Milverton	3058 6790 – 3077 6789	6
5	Old Quarry, Leek Wootton	2896 6914	2.5
6	Driveway, west side of A429 road, Leek Wootton	288 687	4
7	Old Quarry, North Woodloes	276 679 – 276 682	4
8	300 m north of Woodloes Farm	2789 6731	2–3
9	West side of A46 road immediately south of A429 intersection	290 675	11.5
10	Old quarry, Woodloes Estate	2799 6655	11
11	Guy's Cliffe, south-west bank of River Avon	2942 6672 – 2932 6680	8–9
12	East bank of River Avon, Milverton	300 664	6
13	Old quarry, Milverton	301 663	5
14	Coton End Quarry, Warwick	289 655	10–11
15	Shire Hall, Warwick	2800 6502	4
16	The Priory, Warwick	283 653	9
17	Path to Warwick Castle	285 648	5

The base of the formation at Locality 9 rests on the Permian Ashow Formation. There is no basal conglomerate. Beds near the base are also exposed at Locality 10 as follows:

	Thickness m
Sandstone; buff to grey-green, medium-grained, poorly sorted, micaceous, cross-bedded, conglomeratic with pebbles concentrated at bases of foresets; pebbles are mainly sandstone with rare siltstone and quartz; many erosion surfaces	3.4
Sandstone; poorly exposed	6.2
Sandstone (as top unit); irregular base	0.5
Siltstone; grey-green, increasingly red-brown, muddy, micaceous and blocky downwards; passing to	0.32
Mudstone; red-brown, silty, blocky; scattered silt and sand grains	0.28
Sandstone; grey-green, fine-grained, micaceous	0.16
Mudstone; red-brown, silty, locally sandy, micaceous	0.25
Sandstone; grey-green, fine-grained, micaceous	0.05

Some 6 m of basal beds at Locality 4 comprise cross-bedded sandstones containing a persistent 1 m bed of red siltstone. Conglomerates above and below the siltstone contain subrounded pebbles of platy sandstone and siltstone and a few well rounded ones of vein-quartz. The sandstones display ruptured siltstone lenses, and disturbed cross bedding with subvertical foresets.

The middle of the formation is best exposed at Locality 11. The sandstones are generally medium to fine grained, micaceous, and buff to pale grey-green:

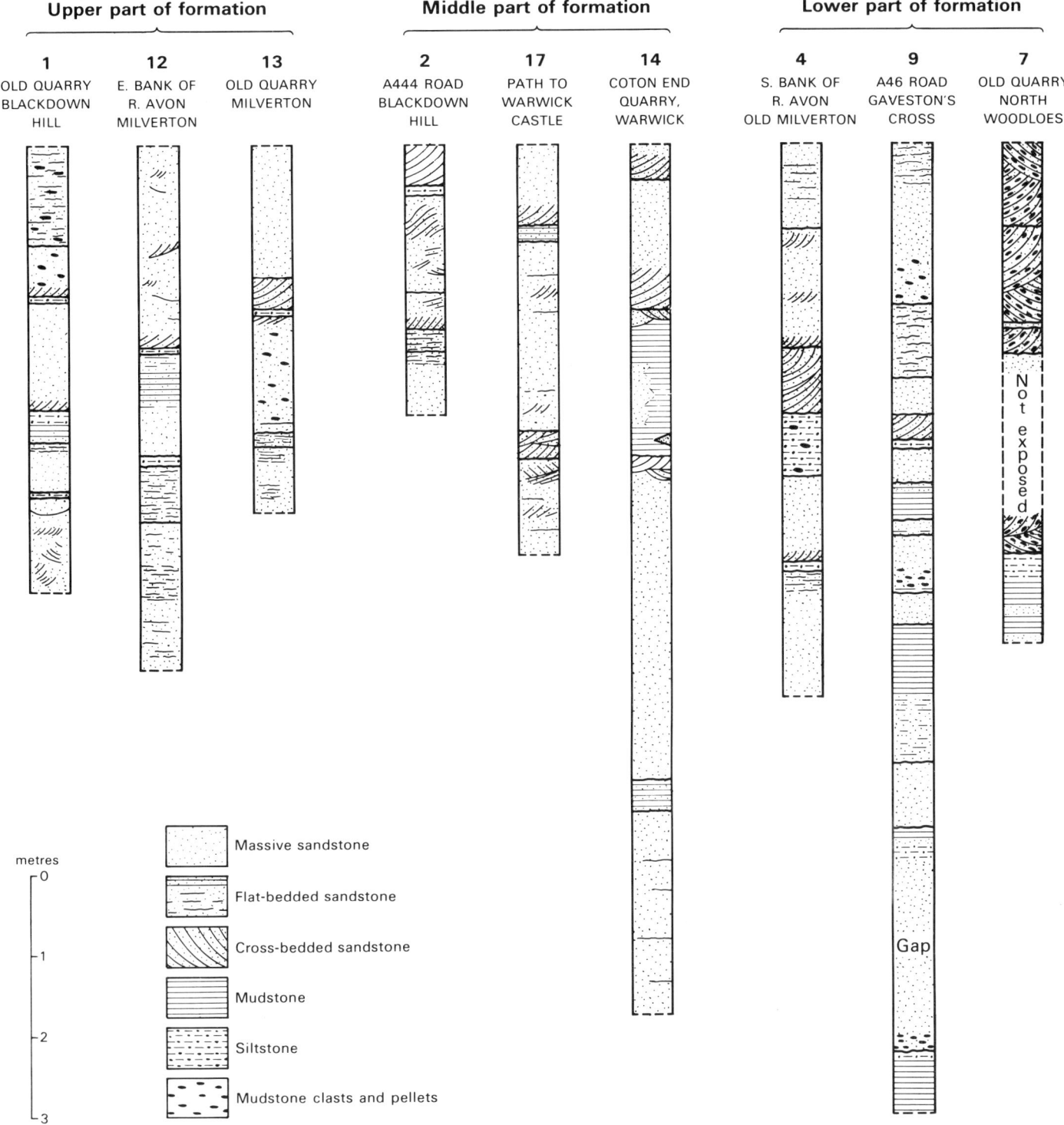

Figure 11   Vertical sections of Bromsgrove Sandstone Formation near Warwick

	Thickness m
Sandstone; with scattered dark brown grains, massive; cross-bedded in basal 0.1 m; erosive base	1.1
Sandstone; locally brown with scattered dark brown grains, strongly cross-bedded; many erosion surfaces; dewatering structure near base; sharp erosive base	0.7 – 1.25

	Thickness m
Sandstone; strongly cross-bedded with scattered sandstone and mudstone clasts; disturbed bedding and dewatering structures in places; erosive base	1.2 – 3.5
Conglomerate; pebbles of sandstone, siltstone and vein-quartz; passes laterally in places into siltstone and mudstone	0 – 0.17
Sandstone; strongly cross-bedded; pinkish-grey calcareous patches 2 to 3 cm in diameter showing	

Figure 12 Principal localities of Bromsgrove Sandstone Formation. See text for list of locality details

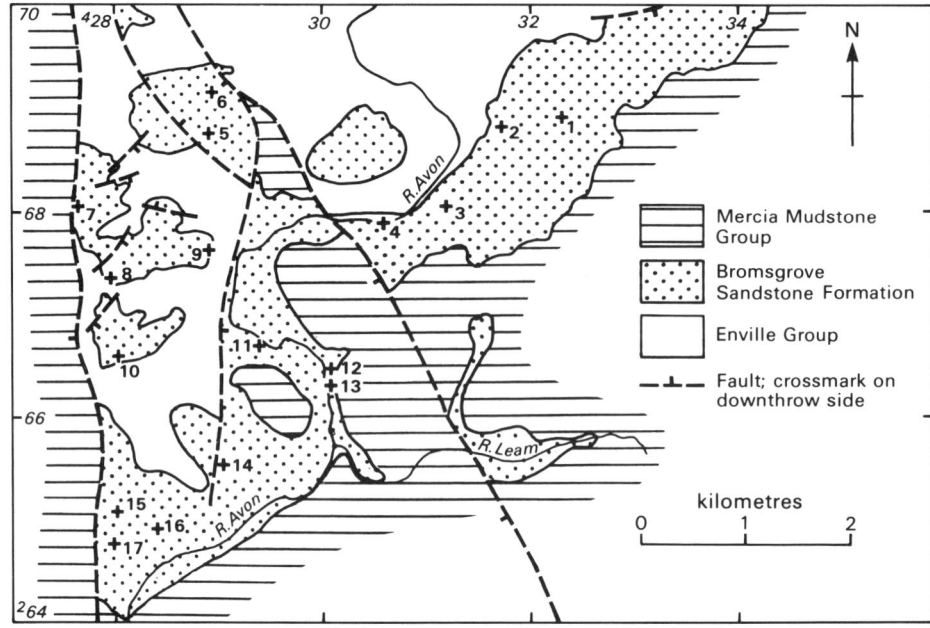

	Thickness m
lustre mottling and commonly elongated parallel to bedding may have formed in a soil horizon; contorted foresets and dewatering structures in places	0.4–1.1
Sandstone; strongly cross-bedded; thinning to north-west	0–0.9
Sandstone; massive to strongly cross-bedded; strong erosion surface separates two cosets	0–1.1
Sandstone; strongly cross-bedded, conglomeratic with platy sandstone fragments; conglomerate pockets at base	0–1.1
Conglomerate; pebbles of calcareous sandstone and vein-quartz; channel lag deposit (Plate 4)	0–0.3
Sandstone; massive and cross-bedded units; scattered calcareous patches	2.15–3.65

At Locality 16, also in the middle beds of the formation, the sandstones are mainly buff and medium to fine grained:

	Thickness m
Sandstone; cross-bedded; erosion surface near top; soft and silty at erosive base	2
Sandstone; generally massive, cross-bedded in part; interbedded with mudstone to east	0.8
Mudstone; red-brown, sandy, silty towards base with a few sandstone laminae; green in basal 0.08 m	0.36
Sandstone; massive, with cross-bedded lenses near top	4
Mudstone; green, silty, micaceous, with interbedded sandstone	0.35
Sandstone; massive, with scattered dark brown grains	1.75

A channel lag-conglomerate fills irregular hollows at the base of a small channel in the higher beds of the formation at Locality 13. It consists of unorientated sandstone pebbles, mainly 1 to 2 cm in diameter but with some up to 12 cm. The bulk of the channel is filled with cross-bedded sandstone. The best section of the topmost beds, at Locality 1, is in medium- to fine-grained micaceous sandstone (Plate 5):

	Thickness m
Sandstone; red-brown to buff, flaggy; lenses of red muddy sand; green mudstone clasts; erosive base	1.3
Sandstone; orange-red to buff, with scattered dark brown grains and green mudstone clasts; cross-bedded in basal 0.1 m; erosive base; wedges out to west	0–0.6
Sandstone; pale greenish grey to buff; generally massive but bedded at top; cross-bedded at base with abundant green siltstone clasts	1.5
Sandstone and siltstone, interbedded; greenish grey, cross-bedded	0.2
Sandstone; yellow-grey to buff, with scattered dark brown grains; mainly cross-bedded with some festoon bedding; dewatering structures; load casts; erosive base	0.6
Sandstone; buff, channelled in part, poorly cross-bedded	0.9–1.5

KA

MERCIA MUDSTONE GROUP

The main outcrop of the Mercia Mudstone follows a broad arc stretching from Princethorpe via Warwick to Burton Green. The strata rest conformably upon the Bromsgrove Sandstone and are overlain by the Penarth Group. Comparative sections across the district are shown in Figure 13.

At outcrop the group is about 135 m thick. It thickens eastwards beneath the Jurassic cover to 179 m in Home Farm Borehole and 205 m in Rugby Waterworks Borehole. West of the Warwick Fault there is a sharp increase in thickness. In Hatton Hospital No. 1 Borehole, 118.5 m were proved, all below the Arden Sandstone, and about 3 km to the west, -just outside the district, Shrewley Borehole, commencing in the Arden Sandstone, proved 196 m.

Plate 4   Channel lag conglomerate and cross-bedded sandstone; Bromsgrove Sandstone Formation, Guy's Cliffe, Warwick.   A13067

The Mercia Mudstone consists predominantly of red-brown blocky mudstones with minor green and laminated mudstones and several thin siltstones and sandstones. The base is gradational, and between Warwick and Cubbington has been taken at a conspicuous green siltstone up to 2 m thick. Beds referable to the 'Waterstones' (Hull, 1869) were not recognised at outcrop, although the basal few metres are commonly silty and micaceous and contain thinly inter-bedded sandstones and mudstones. Richardson (1928) adopted the term 'Passage Beds' for these transitional strata in wells at Royal Leamington Spa, where they are apparently up to 12 m thick. In Home Farm Borehole they were about 6 m thick. Warrington and others (1980, p.39) included most of the 'Waterstones' of the Central Midlands in the Bromsgrove Sandstone, following Wills (1970). However the 'Waterstones' of Nottinghamshire (Elliott, 1961), now the Colwick Formation, were included in the Mercia Mudstone Group by the same authors (1980, pp.51, 57).

Several impersistent thin green siltstones or sandstones (skerries) have been mapped west and south of Warwick in the succeeding red mudstones. A skerry thought to be the Cotgrave Skerry (Elliott, 1961) occurs 55 to 90 m below the base of the Blue Anchor Formation in some boreholes (Figure 13) and crops out north-east of Radford Semele.

The beds above the Cotgrave Skerry proved in Home Farm Borehole are probably referrable to the Edwalton Formation (Elliott, 1961). The top of this formation was taken at the top of the Hollygate Skerry (Elliott, 1961). This is believed to equate with the Arden Sandstone which crops out almost continuously between Harbury and the Princethorpe Fault and comprises a few metres of interbedded coarse grey sandstones and green mudstones. It has not been recognised south of the Whitnash Fault, but may be represented by a thin impersistent skerry south-west of Red House Farm [334 590].

At outcrop the Arden Sandstone lies 15 to 20 m below the base of the Blue Anchor Formation, but in Home Farm Borehole this interval was 39 m (Figure 14). The variation may be due to a minor unconformity at the base of the Blue Anchor Formation, which oversteps a higher skerry near Harbury, or to contemporaneous movement on the Princethorpe Fault. The mudstones between the Arden Sandstone and the Blue Anchor Formation thicken westwards, and the Knowle Borehole [1883 7777] penetrated 108 m below the Blue Anchor Formation without encountering the Arden Sandstone (Old, 1982). These beds have elsewhere been divided into a lower Trent Formation (Elliott, 1961) and a higher Glen Parva Formation (Warrington and others,

Plate 5   Massive sandstone overlying planar-bedded sandstone disturbed by dewatering; Bromsgrove Sandstone Formation, Blackdown Hill.   A13113

1980), but in the present district these subdivisions have been recognised only in Home Farm Borehole.

The Blue Anchor Formation lies at the top of the Mercia Mudstone Group, of which it is the most persistent formation. It consists of 5 to 7 m of pale grey-green silty mudstones with thin siltstones and irregular dolomite concretions.

Macrofossils are extremely scarce throughout the Mercia Mudstone Group and none found are of stratigraphic value. Unidentifiable plant fragments were found in a skerry in Stockton Locks Borehole. In several boreholes the Arden Sandstone contained fish scales, as did the Blue Anchor Formation in Home Farm Borehole. Few of the palynomorph samples examined from the Mercia Mudstone Group in the district were productive. One (MPA 2944), from the Arden Sandstone (at 51.4 m) in Stockton Locks Borehole, yielded cf. *Triadispora obscura*, a miospore indicative of a Ladinian to Carnian age; another (MPA 7012) from the Blue Anchor Formation (at 31.8 m) in the same borehole yielded a sparse miospore association including taxa, such as *Rhaetipollis germanicus*, that are important and typical components of assemblages from the succeeding Penarth Group and are of similar (Rhaetian) age.

Thin sections of rocks from the Mercia Mudstone have been examined by Mr G. E. Strong. The mudstones are fer-ruginous and composed mainly of quartz grains with a little detrital K-feldspar. Gypsum occurs as a cement, and there are some calcite rhombs in scattered crystals and patches. Cream coloured specks in one sample proved to be authigenic K-feldspar (X-ray film X8069). The skerries are calcareous sandstones or siltstones, with angular to subangular quartz and quartzite grains predominant. Calcite cement is ubiquitous, and quartz, gypsum and barytocelestite cements rare. The last occurs as fibrous patches which, in one specimen, had formed later than the calcite.

Arthurton (1980) has likened the depositional environment of the Mercia Mudstone of Cheshire to the present day Ranns of Kutch, an arid coastal flat on the India–Pakistan frontier subject to annual marine inundations and local flooding by fresh water from the adjoining land. This was probably also true of the present district where the red colour, desiccation cracks, evaporites, and the scarcity of fossils in much of the Mercia Mudstone all point to an arid depositional environment, where the detailed section of Home Farm Borehole (Figure 14) shows that most of the strata comprise blocky, unlaminated mudstones, possibly of aeolian origin, and where the skerries were presumably laid down by periodic fresh-water floods. The laminated mudstones are indicative of shallow water, but they are rare ex-

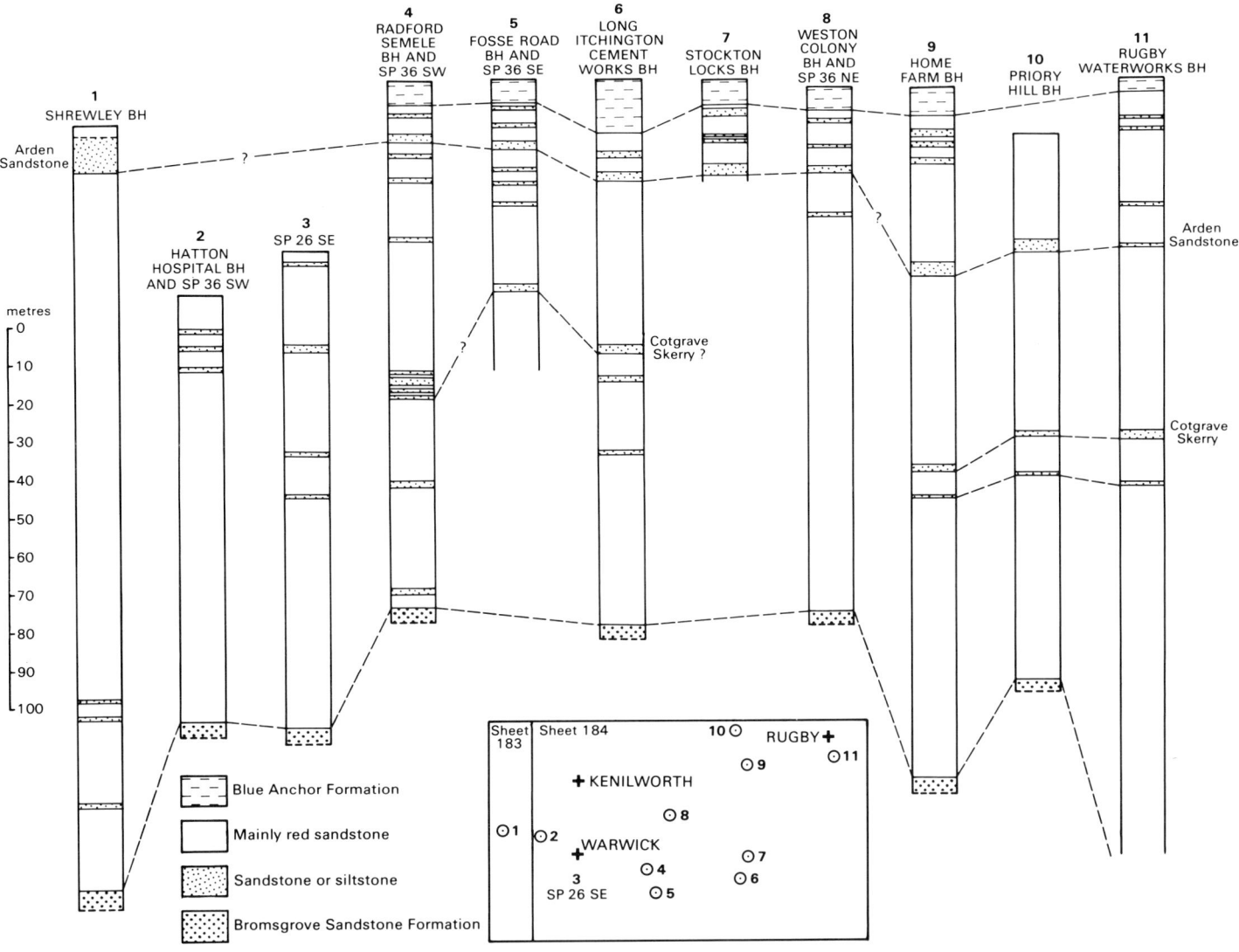

Figure 13 Vertical sections in the Mercia Mudstone Group

cept in the Glen Parva and Blue Anchor formations, which contain no gypsum and have yielded fish scales. It seems likely that towards the end of Norian times the climate of the area ameliorated with the encroachment of the Rhaetian sea.

Details

Home Farm Borehole

Home Farm Borehole provides the only complete section of the Mercia Mudstone Group in the district (Figure 14), and is accordingly described here in considerable detail.

The 'Waterstones' or Colwick Formation (201.2–207.9 m) largely consist of sandy and silty red-brown mudstones with green-grey mottles, together with beds of green-grey, laminated siltstone and very fine-grained sandstone. All are commonly micaceous. Gypsum is present in the mudstones only as a few small blebs and strings. The lower part of the sequence comprises a fining-upwards fluvial cycle, similar to those of the Bromsgrove Sandstone though on a smaller scale; the sandstones are much finer grained than those of the Bromsgrove Sandstone.

The beds between the 'Waterstones' and the Cotgrave Skerry (128.1–201.2 m) have been divided in Nottinghamshire (Elliott, 1961) into the Radcliffe, Carlton and Harlequin formations. At Home Farm the beds largely consist of massive, red-brown mudstones. Darker mudstones below 178.50 m show traces of bedding and lamination, with pale red to cream silt laminae commonly deformed by load structures; they may represent the Radcliffe Formation. At 174.61 m, a thin sandstone bed passes up into silty green-grey mudstone, with a thin layer of contorted laminations at 171.43 m. A tentative correlation is suggested with the Plains Skerry of Elliott (1961), which contains 'slump structures'. In the type area the Plains Skerry occurs about 3 m below the top of the Carlton Formation, a group of mainly massive mudstones, and the beds between 174.61 and 178.50 m may be equivalent to the lower part of this formation. The Harlequin Formation (Elliott, 1961) is characterised by laminated beds. At Home Farm only a few thin laminated layers occur, between 130 and 158 m. Throughout this sequence gypsum, less abundant than in the Edwalton and Trent formations, occurs as rare nodules, tiny blebs and strings; most form sparse subhorizontal fibrous veins, generally more than 1 cm thick. Tiny specks of cream to white, powdery, authigenic

K-feldspar, rarely up to 2 mm across, are locally abundant between 136.09 and 178.50 m; most form coatings on gypsum grains, or partly fill voids presumably left by the solution of gypsum.

In the type area the lower boundary of the Edwalton Formation is the base of the Cotgrave Skerry and the upper, the top of the Hollygate Skerry (Elliott, 1961), regarded as the lateral equivalent of the Arden Sandstone (Warrington, 1970, p.206). In Home Farm Borehole sandstone beds, considered to represent these skerries, suggest that the formation extends from 73.15 to 128.1 m. At 128.1 m, 1.38 m of massive medium-grained sandstone, in parts rather clayey, probably represents the Cotgrave Skerry. The rock is predominantly greenish grey, with red-brown mottles in the more clayey portions, and contains small gypsum nodules. The beds above consist of red-brown, massive mudstones, usually sandy or silty, with ill defined sandy lenticles, scattered coarse sand grains and a few thin clayey sandstones. Gypsum nodules occur sporadically in the upper part of the mudstones, and ramifying veins of fibrous gypsum abounded throughout. Almost all the mudstones are sheared, and break readily into fragments with listric and slickensided surfaces, commonly coated with gypsum. Much of the rock appears to have been fragmented and subsequently recemented; this may reflect solution of primary gypsum nodules and collapse of the rock into the voids thus created, the gypsum being redistributed into veins. Green-grey reduction spots and 'fish-eyes' are associated with the sandier beds, and radioactive grains are presumably scattered throughout the rock. At several horizons 'fish-eyes' are split and displaced by gypsum veins. White and cream specks and streaks of authigenic K-feldspar occur between 108 and 119 m. The basal section of the Arden Sandstone consists of pale grey, medium- to fine-grained sandstone, with a sharply erosive base at 76.7 m, containing many layers of green-grey mudstone clasts. Numerous thin layers and laminae of green-grey mudstone in its upper part are rippled, disrupted by sand-filled polygonal cracks, and penetrated by sandstone load casts. Above 74.14 m the strata consist mainly of sandy green-grey mudstone, with lenticles and patches of sandstone and a few fish scales. At 73.49 m this sandy mudstone passes into smooth purplish brown mudstone containing irregular greyish lilac blotches associated with sandy lenticles. This purple and lilac colouration has been observed elsewhere in this part of the Mercia Mudstone sequence. The top of the Edwalton Formation is taken at the sharp base of a thin, clayey sandstone bed.

The Trent Formation (39.11–73.15 m) consists predominantly of massive, mainly silty, red-brown mudstones with scattered sand grains, poorly defined sandy lenticles, green-grey reduction spots, sporadic 'fish-eyes' and a few very thin sandstones. The one significant skerry (at 48.32 m) consists of 0.93 m of green-grey clayey sandstone split by a bed of red-brown mudstone. A few thin beds of green-grey siltstone, or green-grey mudstone with silt laminae, occur above this skerry. The formation contains abundant veins and nodules of gypsum, the nodules being especially plentiful in the upper part. Beds of pure white gypsum up to 0.3 m thick occur at several levels, roughly corresponding to the commercial gypsum horizons of the Midlands. Cream to pale reddish brown, caliche-like, dolomitic nodules are particularly common in the upper part of the formation, especially in the more gypsiferous mudstones. In places, incomplete dolomitisation had given a brecciated appearance to the mudstone. Thin beds of breccia at 41.70 and 47.31 m, consisting of fragments of red-brown mudstone in a darker matrix, resemble the 'vein type' breccias of Elliott (1961, p.206). Between 67.99 and 68.44 m, rock composed of very small horizontally aligned elongated fragments of red-brown mudstone in a dark green matrix may be a 'flow-type' breccia (Elliott, 1961, pp.205, 206).

Following Elliott (1961, p.222), the base of the Glen Parva Formation is taken beneath a green-grey, poorly laminated mudstone containing scattered fish scales, and the formation is taken to extend

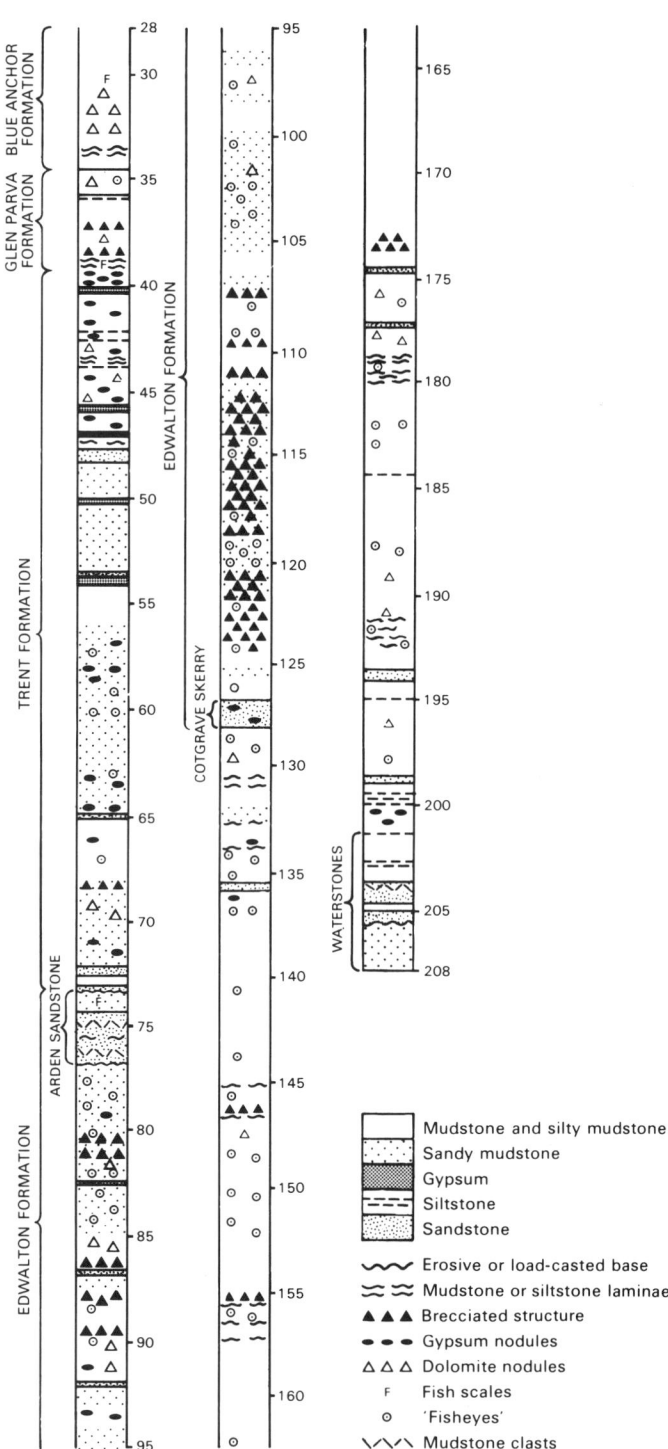

Figure 14   The Mercia Mudstone in Home Farm Borehole

from 34.49 to 39.11 m. The strata above the basal mudstone comprise red-brown and green-grey mottled, poorly bedded or massive mudstones with subordinate pale green-grey mudstone and siltstone. Cream-coloured, caliche-like, dolomitic nodules are common throughout, and dolomitic patches occur in places. Gypsum is entirely absent. Flow-type breccias occur from 36.99 to 37.30 m and from 38.11 to 38.55 m.

The Blue Anchor Formation (27.92 – 34.49 m) consists of pale green-grey, silty mudstone with a few thin beds of green-grey dolomitic siltstone. Cream to grey, caliche-like, dolomitic nodules and poorly defined dolomitic patches are common throughout. Scattered fish scales are present in the upper part of the formation, and a few small blebs and nodules of pyrite occur in the uppermost 3 m. The topmost 0.04 m is brecciated, the spaces between fragments being filled with dark grey mudstone of the overlying Westbury Formation. Brecciation appears to have occurred in situ, perhaps as a result of desiccation (cf. Kent, 1953, p.135).   MGS

Ryton-on-Dunsmore to Princethorpe

Reddish purple mudstone with subordinate green mudstone and siltstone crops out in the bottom of a gravel pit [3928 7367] south-west of Grange Farm; the strata resemble the top beds of the Arden Sandstone of Home Farm Borehole. Other indications of the Arden Sandstone are in Ryton No. 12 Borehole [3928 7386] where 1.5 m of red-brown, lilac and grey marl and silty marl overlay 4.6 m of green-grey marl and silty marl at 12.2 m; in Ryton No. 6 Borehole [3889 7362] which proved 3.0 m of grey sandy marl beneath gravel at 10.7 m; in Ryton No. 7 Borehole [3906 7357] where 3.7 m of green marl with grey sandstone bands lay at 15.5 m; in trial boreholes for sand and gravel [386 733 to 389 738] where green clay directly underlay the Baginton Sand and Gravel; in Ryton No. 5 [3916 7431] and 5a [3915 7430] boreholes, which showed traces of grey marl and pale grey sandstone beneath gravel; in a nearby road cutting [3914 7418] where green-grey silty clay was augered at the base of the Baginton Sand and Gravel; and in shallow boreholes east and south-east of Grange Farm [395 740] which proved green and grey clay. Mudstone probably belonging to the Glen Parva Formation and the upper part of the Trent Formation is exposed at Stretton-on-Dunsmore [408 727]. Gypsum was once mined there (Chapter 8). An outlier of Arden Sandstone at Windmill Hill [408 700] is marked by brash of grey-green, medium- to coarse-grained dolomitic sandstone with pseudomorphs after halite and mudstone flakes.   KA, MGS

Royal Leamington Spa to Ufton

Details of wells in Royal Leamington Spa were given by Richardson (1928, pp.114 – 120). Shotton (1929, p.189) recorded 7 m of thinly interbedded red and green mudstone and sandstone close to the base of the Mercia Mudstone in a quarry [303 663] (now filled) at Old Milverton Road. In Weston Colony Borehole [3662 6907] red marl rested directly on Bromsgrove Sandstone at 51.8 m (Butler, 1946, p.44). Some 10 m of red, silty, micaceous mudstone with green reduction spots, exposed [325 664] SW of Lillington church, lie about 25 m above the base of the Mercia Mudstone Group. Strata at about the same level include 3.5 m of red mudstone with green reduction spots at Glebe Farm [339 674], and red, silty, micaceous mudstone, brecciated in parts, with green silty reduction spots and a few green siltstone nodules, in an old pit [337 664] to the SSW.

The Cotgrave Skerry crops out [3503 6487] as 1 m of grey-green sandstone with red and green mudstone lenses and partings, overlain by 0.25 m of red mudstone with wind-polished sand grains, NW of Leasowe Farm . It may also be represented by 1.8 m of grey sandstone at 14.0 m in Fosse Road Borehole [3530 6249], and 1.5 m of green-grey, shaly, calcareous sandstone with thin partings of sandy mudstone at 91.1 m in Long Itchington Cement Works Borehole [4199 6396] (Woodland, 1942, p.6) (Figure 13). A thin sandstone exposed in a ditch [390 686] near Eathorpe lies about 25 m above the Cotgrave Skerry:

	Thickness m
Mudstone; red with green reduction spots	0.5
Sandstone; mainly red and green with sandy mudstone clasts; dark yellow to red-brown and coarser grained in lower part; thin layers of green mudstone near base	1.0
Mudstone; red	0.2

A siltstone seen in a trench [3689 6295] NW of Ufton lies about 30 m above the Cotgrave Skerry:

	Thickness m
Mudstone; red-brown, silty, with green reduction spots; passing to	0.15
Mudstone; green, silty, with irregular top	0.35
Siltstone; grey-green, poorly fissile, dolomitic; sporadic green mudstone clasts; fine-grained sandstone beds; scattered wind-polished sand grains and quartz pebbles up to 3 mm in diameter; some malachite mineralisation	0.20
Mudstone; red-brown, silty, with green reduction spots; irregular 0.05 m green reduction zone at top	0.30

In Stockton Locks Borehole (Figure 13) the Glen Parva Formation was absent. The Trent Formation was only 13.5 m thick; its lower half was gypsiferous and included a 0.2 m bed of gypsum at 43.66 m. Purple mudstone occurred between 48.7 and 49.5 m. The Arden Sandstone (2.8 m thick at 52.28 m) was mainly green mudstone, with thin layers and lenses of coarse- and fine-grained sandstone, siltstone and dolomitic mudstone. Sand-filled burrows and mudcracks were common, some cutting across dolomitic layers. Other sedimentary structures included collapsed burrows, sediment balls and small-scale cross-bedding. Between 50.95 and 51.22 m red mudstone with sandy laminae and burrows rested on interlayered red and green mudstone overlying 3 cm of pebbly sandstone with an erosive base.

The Arden Sandstone of the Ufton area contains more sandstone than that in Stockton Locks Borehole; temporary exposures showed interbedded sandstone and mudstone in beds usually about 0.2 m thick. Farther west, mudstone predominates in a section [3707 6447] SE of Bunkers Hill Farm:

	Thickness m
Mudstone; red-green mottled	0.20
Mudstone and sandstone; green, interbedded	0.08
Sandstone;grey-green, medium- to fine-grained, cross-bedded; passing laterally into mudstone	0.05
Mudstone; grey-green, silty	0.52
Mudstone; red, green-mottled	0.30

An outlier of Arden Sandstone at Highdown Hill Plantation [330 610] comprises 5.5 m of green mudstone with thin ribs of sandstone and a lenticular red mudstone.   JB

Thin sandstones above the Arden Sandstone are exposed in the railway cutting [3822 6654] north of Snowford Lodge:

	Thickness m
Mudstone; red	1.0
Sandstone; pale green, medium-grained with scattered red pebbles 2 mm in diameter; passing to	0.3
Mudstone; green	0.3
Sandstone; soft, grey-green, medium- to fine-grained	0.2

	Thickness m
Mudstone; red	0.2
Mudstone; red, with gypsiferous and sandy layers	1.3
Mudstone; green, red-mottled near top; gypsiferous in parts; grey-green sandstone layers up to 0.05 m thick	0.9
Mudstone; red, green spots in top 0.2 m; sandy in parts	1.3
Sandstone; fine-grained, green	0.04
Mudstone; red, with green spots and gypsum veins	2.0

The Blue Anchor Formation of Stockton Locks and Harbury Quarry boreholes resembled that proved in Home Farm Borehole (Figure 14). In the former the base of the formation was irregular and marked by scattered mudstone pebbles and 'fish eyes'. In Harbury Quarry Borehole 1 m of medium-grained sandstone occurred 1 m above the base of the formation. The basal 2.7 m of the Blue Anchor Formation in an old brick-pit [4142 6587] north of Long Itchington comprises grey-green mudstone resting on 0.2 m of interbedded grey-green and purple mudstone passing down into red mudstone.

Budbrooke to Wasperton

Mercia Mudstone strata west of the Warwick Fault lie below the Arden Sandstone. Hatton Mental Hospital No. 1 and No. 2 boreholes [2478 6715] (Richardson and Fleet, 1926; Butler, 1946, p.45) proved 5.0 and 5.6 m respectively of basal passage beds (p.27). The higher of two skerries mapped hereabouts crops out in a pond [2500 6789], and in a ditch 100 m to the north-east, as 0.1 m of greenish grey, medium-grained sandstone in red mudstone. Fragmentary green siltstone and mudstone in red mudstone in a ditch [2723 6698] belong to the lower skerry. Skerries west and east of Wasperton are at a similar stratigraphic level. At Scar Bank [2565 5883], on the west side of the River Avon, a section shows:

	Thickness m
Mudstone; red-brown, silty; many wind-polished sand grains; rare small pebbles; green mottles and spots; lenses of green and red-brown mottled sandstone and siltstone up to 0.07 m thick	5 – 6
Siltstone; green-grey, blocky; scattered wind-polished sand grains; fissile near top with green mudstone partings and lenses; rare quartz pebbles 2 – 3 mm across; passing into	0.35
Mudstone; red-brown, blocky; sandy in top 0.5 m; sporadic sandy pockets; scattered wind-polished sand grains	0.7

The exposure is cut by a small near-vertical fault downthrowing about 6 m ESE. A further 3 m of the lowest bed are exposed on the upthrow side of the fault.   KA

PENARTH GROUP

The Penarth Group characteristically forms a NW-facing escarpment, a westerly extension of which near Print Wood [386 649] is attributable to trough faulting. Landslips, lubricated by springs at the base of the Langport Member, make it impossible to delineate the lower subdivisions, and in places obscure the base of the group. Typical thicknesses (m) are:

	Rugby Waterworks Borehole	Home Farm Borehole	Stockton Locks Borehole	Harbury Quarry Borehole
Langport Member	3.7	1.98	1.97	2.55
Cotham Member	12.2	11.77	11.46	9.46
Westbury Formation	8.5	4.82	6.20	7.06
Total	24.4	18.57	19.63	19.07

The Westbury Formation generally rests conformably upon the Blue Anchor Formation, although signs of local erosion have been observed at the contact, as in Harbury Quarry Borehole and Stockton Locks Borehole. It consists of dark grey to black, fissile mudstone with abundant lenticles and laminae of white micaceous silt and fine sand, commonly exhibiting small load structures, interbedded with layers of more massive mudstone in which silty bands are less common. The dark colour of the mudstones is partly attributable to disseminated pyrite, and blebs and nodules of the mineral are plentiful. The junction with the Cotham Member is gradational.

The Cotham Member consists of pale to medium green-grey, smooth, blocky, calcareous mudstone with some olive-brown layers and mottles. Lenticles of laminated white silt, abundant in places, show small-scale cross-bedding, slumping, convolute bedding and flame structures. The lower beds are darker grey, less calcareous and more silty, with a few calcareous nodules; they pass down into grey shales ('Transition Beds' of Orbell, 1973).

The Langport Member comprises hard, porcellanous, massive limestone, pale grey when fresh but weathering to white or cream with a yellow tinge attributable to weathered pyrite, with a few thin partings of dark grey mudstone. At outcrop, the rock is generally worn and pitted by solution. The base of the member is everywhere sharp; its top is commonly bored, eroded and, in places, conglomeratic.

The Westbury Formation proved in the boreholes yielded a marine, bivalve dominated macrofauna which includes *Cardinia regularis, Eotrapezium concentricum, E. sp., 'Gervillia' praecursor, Modiolus* cf. *hillanoides, Mytilus sp., Protocardia rhaetica, Rhaetavicula contorta* and rare ophiuroids. Silty and sandy lenticles, particularly near the base, yield *Acrodus, Gyrolepis, Hybodus* and other fish fragments. The fauna is typical of that of the *Rhaetavicula contorta* Zone of the Rhaetian (Warrington and others, 1980, p.18).

The Cotham Member has a sparse macrofauna with *Dimyopsis [Dimyodon] intusstriatus, Liostrea* and fish fragments from Harbury Quarry Borehole and sporadically an abundance of the small branchiopod *Euestheria minuta* in the other boreholes. The white limestones of the Langport Member are rarely rich in macrofossils; *Dimyopsis, Liostrea hisingeri, Modiolus* and echinoid fragments occur. Richardson (1912, p.31) recorded from the White Lias of Harbury railway cutting the coral *Montlivaltia tomesi,* bivalves including *Eotrapezium [Isocyprina], Dimyopsis [Dimyodon] intusstriatus, Meleagrinella [Pseudomonotis], Modiolus [Volsella] minimus, Plagiostoma valoniensis, Plicatula* cf. *cloacina, Pteromya [Pleuromya] crowcombeia* and ostracods. *Hemipedina [Pseudopedina] tomesi* is also included. Echinoid spines from these beds are usually now referred to

Diademopsis but a *H. tomesi* from Print Hill, in Warwick Museum, was regarded by Woodward (1893, p.160) as evidence that this hill contained Lower Lias which was not proved in the present survey. Print Hill also yielded the specimen of *Montlivaltia rhaetica* described by Tomes (1878). Records of fossils from the 'White Lias' of King's Newnham and Church Lawford (Oldham and Jones, 1879) contain a variety of taxa which cannot be evaluated without verifying the specimens.

Dr G. Warrington reports palynomorph assemblages from the Penarth Group of Stockton Locks Borehole (Figure 15) are more profuse and varied than those from the Blue Anchor Formation; they are indicative of a marine environment and are assigned a Rhaetian age. Assemblages from the Westbury Formation are dominated by miospores, principally *Classopollis torosus*, *Ovalipollis pseudoalatus*, *Rhaetipollis germanicus* and *Ricciisporites tuberculatus*, associated with organic-walled microplankton, principally the dinoflagellate cyst *Rhaetogonyaulax rhaetica*, but including acritarchs and sporadic Tasmanaceae. Assemblages from the Cotham Member are more varied. In the lower part dinoflagellate cysts are abundant, and dominant at some levels; but they are less abundant in the upper part, the assemblages in these beds and in the Langport Member being dominated by miospores. A marked change in the composition and a reduction in the diversity of the miospore associations occur at the top of the Cotham Member, and an assemblage from the Langport Member is dominated by *Kraeuselisporites reissingeri*.

The appearance of foraminifera at the base of the Westbury Formation marks the advent of a regional marine transgression, the progress of which is reflected in the macro- and microfossil associations of the Penarth Group. The dark grey pyritous, argillaceous beds of the Westbury Formation formed under reducing conditions in shallow water where current activity was sufficient to distribute influxes of arenaceous sediments only as lenticles and thin beds. The pale grey, calcareous beds of the Cotham Member contain few macrofossils and are commonly regarded as having formed in shallow bodies of fresh or brackish water. However, the organic-walled microplankton components of palynomorph assemblages from the lower beds of the member are comparable with those from the underlying Westbury Formation and are associated with remains of foraminifera. A marine influence is therefore likely in at least the lower half of the Cotham Member. The presence of corals and only local erosion and desiccation surfaces suggest that the pure, porcellanous limestones of the Langport Member were probably chemically precipitated from a shallow, warm and normally clear sea subject to only minor influxes of argillaceous sediment.

Details

Princethorpe to Rugby

In Home Farm Borehole the basal 2 cm of the Westbury Formation contained scattered fish and bone fragments. The brecciated top of the underlying Blue Anchor Formation is cemented with dark grey mudstone. A thin bed of slumped, isoclinally folded mudstone near the top of the Westbury Formation resembles strata common elsewhere near the base of the Cotham Member (Poole, 1969, p.14). In the Church Lawford railway cutting [443 759] Woodward

(1893, p.162) noted 1.5 to 2.4 m of greenish-grey marl overlain by limestone (Langport Member, see below) and resting on red and green marl. He regarded the last as 'Keuper', and stated that the black paper shales (now Westbury Formation) were absent. Both the Cotham Member and the Westbury Formation were, however, proved by augering at this locality during the recent survey, and both the 'greenish-grey marl' and the 'red and green marl' probably belong to the Cotham Member. Reddened mudstones in the Cotham Member also occur at Rugby, and in north Lincolnshire and south-east Yorkshire (Kent, 1968, p.178).

Green-grey, fissile clays of the Cotham member are exposed in the bed of the River Avon [4560 7618] east of Church Lawford and in the river bank [4601 7723] at King's Newnham where 3 m of green-grey mudstone lie beneath cryoturbated limestones of the Langport Member. The mudstone is mainly blocky, with conchoidal fracture, but locally fissile with wisps and laminae of white silt; in the weathered zone beneath the limestones the mudstones contain small pieces of race. The number 4 Borehole [4920 7571] of the Rugby Portland Cement Co. proved 4.8 m of interbedded green shale and red shale beneath the Langport Member.

Subdivisions of the Penarth Group and underlying strata proved in Rugby Waterworks Borehole [5078 7375] (Wilson, 1869; Woodward, 1893; Richardson, 1912, 1928; Clements, 1977) are now classified as follows (Richardson's beds 13 and 14 were apparently transposed in error):

	Bed No. (Richardson, 1928)	Lithological record	Thickness m
Langport Member	11	Stone	3.7
Cotham Member	12	Light blue clay	0.6
	14	Brown clay	11.6
Westbury Formation	13	Dark rotten clay	6.1
	15	Very black clay	2.4
Blue Anchor Formation	16	Very light hard stone	3.0

Similar strata in a borehole in Rugby, [5100 7646] of the British Thomson-Houston Company were:

	Bed No. (Richardson, 1928)	Lithological record	Thickness m
Langport Member	6	Rock and cementstone	4.4
Cotham Member and Westbury Formation	7	Brown clay	5.2
	8	Rock	1.5
	9	Blue and brown clay	7.5
	10	Rock	4.1
	11	Black clay	0.6
Blue Anchor Formation	12	Rock	3.8
	13	Clay and rock	0.6

Richardson (1928) regarded beds 10 and 11 as '?Rhaetic' and the beds above as Lower Lias. Clements (1977, p.39) classified bed 10 as 'White Lias' and bed 11 as 'Rhaetic Black shales'.

The Langport Member in the Church Lawford railway cutting [4436 7596] comprised 2.6 m of massive, but locally laminated, fine-grained, grey-white limestone, fissured and veined with calcite; the top of the bed was decalcified and ferruginous, and pitted by

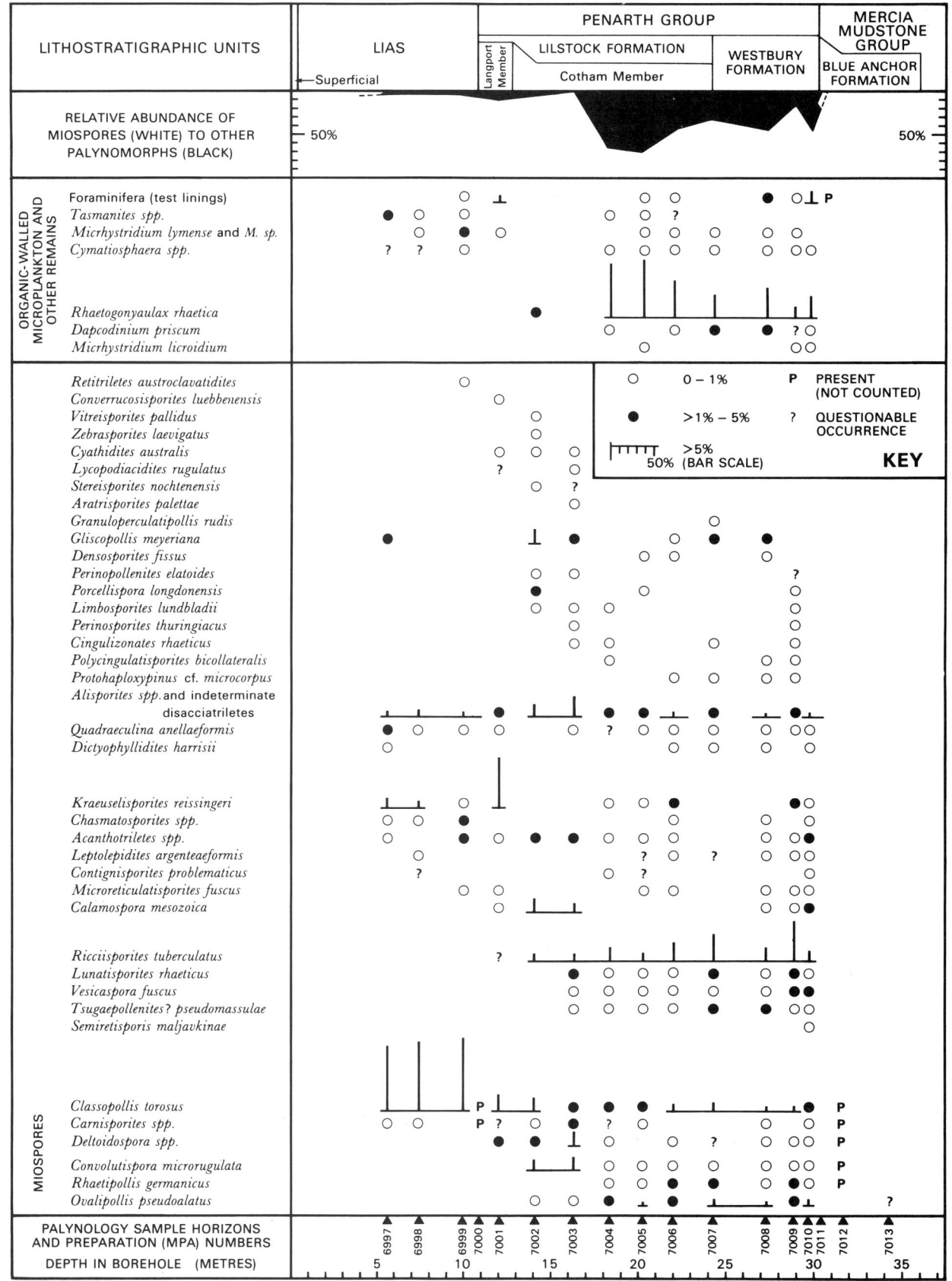

Figure 15 Stratigraphical distribution and relative abundances of palynomorphs from the Mercia Mudstone Group to Lias succession of Stockton Locks Borehole

solution. Woodward (1893, p.192) recorded 1.5 to 1.8 m of limestone there. Limestones cause rapids in the River Avon southeast of King's Newnham [4587 7603: 4616 7632; 4633 7725]. Debris from a small reservoir [4614 7704] includes large blocks of hard, recrystallised brown-grey limestone which contain scattered birdseyes and shell debris including recrystallised bivalves and echinoid spines. The limestone is commonly deeply pitted by solution, and characterised by a powdery yellowish white decalcified coating. Newnham Regis Baths (Bailey, 1582; Richardson, 1928, pp.51, 52) [?4597 7728] were fed by a spring rising from Langport Member limestones. Several springs near Bath Cottage [4619 7729] are marked by tufa, and 'petrifying springs' rising from limestones at Stretton-on-Dunsmore and Limestone Hall [442 754] were mentioned by Oldham and Jones (1879).   MGS

Birdingbury to Harbury

The railway cutting [4276 6884] at Birdingbury shows the basal 1.5 m of the Langport Member to comprise buff or pale grey, recrystallised limestone; borings 0.3 m above the base are filled with orange-brown clay and sand. The Langport Member of Stockton Locks Borehole consisted of 1.97 m of hard, pale to medium grey, fine-grained limestone at 12.84 m. It has an irregular eroded upper surface, many rippled mudstone partings and wisps up to 0.03 m thick, and rare bored surfaces. A thin layer of conglomerate at 12.05 m comprised pebbles (up to 0.01 m) set in sand. The fauna was sparse, with scattered bivalves and shell debris in some mudstone partings. Calcite and pyrite veins occurred throughout. An almost identical sequence was proved in Harbury Quarry Borehole.

A ditch [3973 6300] at the Bascote Heath crossroads shows 2 m of pale grey-green mudstones of the Cotham Member; lenticles of pale grey, fine-grained, sandy, rippled limestone are present, together with a few nodules of pale grey, slightly sandy porcellanous limestone.

A borehole [4202 6302] at Long Itchington Quarry proved 0.15 m of limestone 1.7 m below the top of the Cotham Member. Richardson (1912, p.31) described the top of the White Lias (Langport Member) in Harbury railway cutting [370 607] as iron-stained and crowded with fossils, particularly *Isocyprina* (now *Eotrapezium*). Exposures in the Cotham Member in a road cutting [344 604] at Blaken Knob comprise 2 m of calcareous clay with silty laminae, and with layers of chocolate-brown clay up to 0.5 m thick.

JB

CHAPTER 6

Jurassic

Early Jurassic strata crop out over the eastern part of the district where a substantial Liassic sequence is capped, in the extreme south-east, by Northampton Sand. The lithostratigraphic classification follows Cope and others (1980). The Blue Lias is facies controlled, and the biostratigraphic position of its base varies within the district. The same is probably true of the Lower–Middle Lias boundary. Litho- and biostratigraphical correlations are shown in Table 3. The ammonite zones and subzones follow Dean and others (1961), except that the *Echioceras raricostatum* Subzone is replaced by *E. raricostatoides* (Getty, 1973) and that there have been subzonal changes in the *Coroniceras bucklandi* and *Arnioceras semicostatum* zones following Ivimey-Cook and Donovan *in* Whittaker and Green (1983). Approximate zone thicknesses are given in Table 3, and Figure 16 shows conjectural zone outcrops for the Lower Lias.

LOWER LIAS

Lower Liassic rocks occupy virtually the entire Jurassic outcrop (Figure 1). The generally NE-trending outcrop is affected by several faults. The Princethorpe Fault displaces the Lias outcrop by up to 5 km; at Harbury a complex of small faults disturbs the lowest beds; a fault-bounded trough of Blue Lias and underlying mudstones extends north-westwards to Whitnash; and at Napton on the Hill faulting is inferred from borehole information, the height of the Middle Lias outcrops, and the thicknesses of Lower Lias.

The Lower Lias rests unconformably on the Langport Member. As it is predominately mudstone it forms fairly flat terrain except for the low Blue Lias escarpment and minor limestone features. In the east and south-east the topography rises substantially up to the Marlstone Rock Bed-capped escarpment. The total thickness of the Lower Lias in the district is 200–220 m. This compares with 150–175 m in the Market Harborough district (Poole and others, 1968); 230–250 m in the Vale of Belvoir (information from confidential NCB boreholes); and about 200 m in the Northampton area (Thompson, 1880). Thicknesses are variable in the Banbury (Edmonds and others, 1965) and Stratford upon Avon (Williams and Whittaker, 1974) districts, but generally lessen towards the London Platform.

MUDSTONES BELOW THE BLUE LIAS

Grey, blocky to fissile, shelly, often calcareous, mudstones at the base of the Lias contain a few grey, fine-grained cementstones as impersistent layers and nodule bands. At surface the mudstones weather to dark greenish grey clay. Thicknesses at or near outcrop range from about 11 to 20 m, being greatest in north-west Rugby. South-eastwards the mud-

stones thin, the Gas Council Napton No. 1 [4494 6116] and No. 2 [4623 6061] boreholes proving 6.1 and 1.9 m respectively. The beds are absent 15 km to the east in NCB Hollowell Borehole [6833 7183], and to the south in the Withycombe Farm Borehole (Poole, 1978) near Banbury. Variations in thickness within the district are due largely to lateral passage of the uppermost beds into the Blue Lias, and to overlap of the lowest beds (see below).

To the north-east of the Warwick district the basal mudstones and limestones of the Lias are of late Triassic age (the pre-*planorbis* Beds of Richardson), but within the district these beds are not proved. The Church Lawford railway cutting [444 759] is thought to contain beds of *planorbis* Zone age resting directly on the Langport Member (Arkell, 1933, p.135). Abundant *Psiloceras planorbis* occur between 8.11 to 8.28 m in Home Farm Borehole [4317 7309], Stretton on Dunsmore (Sumbler, 1980), here the underlying 1.07 m of mudstones are also assigned to this zone. Further south there is no evidence for the *planorbis* Zone at Southam or Harbury. At Southam *Laqueoceras* and *Waehneroceras* establish the presence of the upper part of the *liasicus* Zone close above the Langport Member. The Harbury Quarry Borehole [3922 5889] (Brewster, 1978) penetrated 13.24 m of Lias mudstones below the Blue Lias with the highest 9.98 m of these yielding evidence of the *liasicus* Zone with schlotheimiid ammonites; the *planorbis* Zone was not proved. Beyond the district to the south, Withycombe Farm Borehole, Banbury (Poole, 1978) proved thin *angulata* Zone limestones on Langport Member as the earlier beds of the Lias are overlapped south-eastwards onto the northern flank of the London Platform (Donovan and others, 1979).

The age of the top of the mudstones also youngs southward. At Rugby the mudstones extend into the highest subzone of the *liasicus* Zone (Clements, 1977). In Harbury Quarry Borehole an indeterminate schlotheimiid at 17.34 m, just below the Blue Lias, may indicate the *angulata* Zone. The *angulata* Zone was proved below the Blue Lias at Birdingbury, and at Long Itchington the junction of the *angulata* and *liasicus* zones probably lies just above the base of the Blue Lias (Clements, 1975). Faunas are generally sparse. Ammonites present include *Psiloceras*, *Laqueoceras*, *Schlotheimia* and *Waehneroceras*. Bivalves include *Anningella*, *Astarte*, *Cardinia*, *Chlamys*, *Liostrea*, *Lucina*, *Modiolus*, *Oxytoma?*, *Parallelodon*, *Plagiostoma*, and *Pseudolimea*. The trace fossils *Chondrites* and *Diplocraterion*, indeterminate pyritic trails, crinoid columnals, echinoid spines, foraminifera, ostracods and fish fragments are also present.

Palynomorph assemblages from the Lias in Stockton Locks Borehole are of limited diversity; they are dominated by miospores, especially *Classopollis torosus* and small numbers of *Kraeuselisporites reissingeri*, and are similar to those from the underlying Langport Member. The organic-walled microplankton associations comprise small numbers of acanthomorph and herkomorph acritarchs and Tasman-

Table 3 Correlation of Jurassic chronostratigraphy, biostratigraphy and lithostratigraphy for the Warwick district

Stage	Ammonite Zone	Subzone	Lithostratigraphy		Thickness (m)
AALENIAN	*Leioceras opalinum*	*Tmetoceras scissum*	**Northampton Sand Formation**	Ferruginous sandstone	c.2–5
		Leioceras opalinum			
TOARCIAN	*Dumortieria levesquei*	*Pleydellia aalensis*			
		Dumortieria moorei			
		Dumortieria levesquei			
		Phlyseogrammoceras dispansum			
	Grammoceras thouarsense	*Pseudogrammoceras fallaciosum*			
		Grammoceras striatulum			
	Haugia variabilis				
	Hildoceras bifrons * 30+	*Catacoeloceras crassum*	~~~~~~~~~~~~~~~~~~~~~~~~~~~~~~~~		
		Peronoceras fibulatum			
		Dactylioceras commune	Upper Lias	Grey mudstone	34–38
	Harpoceras falciferum * 3–4	*Harpoceras falciferum* *			
		Harpoceras exaratum *		Limestones and shales, thin	—
	Dactylioceras tenuicostatum * 0.15	*Dactylioceras semicelatum*	Transition Bed	Oolitic ferruginous limestone	c.0.05
		Dactylioceras tenuicostatum			
		Dactylioceras clevelandicum			
		Protogrammoceras paltum			
PLIENS-BACHIAN	UPPER	*Pleuroceras spinatum* 2.5–4	*Pleuroceras hawskerense*	Marlstone Rock Bed	Ferruginous limestones, sandstones and oolites	2.5–4
		Pleuroceras apyrenum			
	Amaltheus margaritatus * 10–35	*Amaltheus gibbosus* *	Middle Lias Silts and Clays		
		Amaltheus subnodosus *			
		Amaltheus stokesi *		Grey siltstones, sandstones and mudstones, commonly ferruginous	10–35
	LOWER	*Prodactylioceras daveoi* * >18.6	*Oistoceras figulinum* *	Lower Lias	Grey mudstone with ironstone and cementstone nodules	c.30–?40
		Aegoceras capricornus			
		Aegoceras maculatum *	'85' Marker Member	Shell detrital limestone and calcareous mudstone	2–3
	Tragophylloceras ibex * ?15–49.8	*Beaniceras luridum* *			
		Acanthopleuroceras valdani *			
		Tropidoceras masseanum *		Grey mudstone with cementstone nodules	25–50
	Uptonia jamesoni * 3.65	*Uptonia jamesoni*			
		Platypleuroceras brevispina *	'70' Marker Member	Very shelly marl and calcareous mudstone with limestone nodules	up to 4.5
		Polymorphites polymorphus			
		Phricodoceras taylori		Grey mudstones	

Table 3 *continued*

Stage		Ammonite Zone	Subzone	Lithostratigraphy		Thickness (m)
	UPPER	*Echioceras raricostatum** 27.25	*Paltechioceras aplanatum**			
			*Leptechioceras macdonnelli**			
			Echioceras raricostatoides			
			*Crucilobiceras densinodulum**			
SINE-MURIAN		*Oxynoticeras oxynotum* >9	*Oxynoticeras oxynotum*	Lower Lias	Grey mudstone with occasional cementstones	85–100
			Oxynoticeras simpsoni			
		Asteroceras obtusum	*Eparietites denotatus*			
			Asteroceras stellare			
			Asteroceras obtusum			
	LOWER	*Caenisites turneri** 40–50	*Microderoceras birchi**			
			*Caenisites brooki**			
		*Arnioceras semicostatum**	*Euagassiceras resupinatum*			
			Agassiceras scipionianum			
			Coroniceras lyra			
		*Coroniceras bucklandi**	*Coroniceras rotiforme**	Blue Lias	Alternating grey mudstones and cementstones with thin paper shales	25–40
			*Vermiceras conybeari**			
HETTANGIAN		*Schlotheimia angulata** 12–15	*Schlotheimia complanata*	Calcirhynchia calcaria Bed		
			Schlotheimia extranodosa			
		*Alsatites liasicus** c.10	*Alsatites laqueus**	Lower Lias	Grey mudstone with a few cementstones	11–20
			Waehneroceras portlocki			
		*Psiloceras planorbis** 3.35	*Caloceras johnstoni*			
			Psiloceras planorbis			
RHAETIAN				Penarth Group, Langport Member		

Ammonite zones and subzones proved in the area are indicated (*). Estimates of the thickness of beds representing the zones are in metres.

The *daveoi* to *oxynotum* zones were proved in the Barby Borehole and the *planorbis* Zone in the Home Farm Borehole.

aceae. Dinoflagellate cysts, a major component of many assemblages from the Penarth Group, are absent. Remains of foraminifera occur sporadically (Figure 15).

The absence of coarse sediment in the *planorbis* and *liasicus* zones, coupled with a mixed benthonic and free-swimming fauna, suggests quiet, aerated tropical waters. The fauna is sparse compared with that of later Liassic times. Scattered cementstones point to the occasional presence of warm, fairly shallow water receiving little terrigenous material.

Details

Harbury to Leamington Hastings

The Harbury Quarry Borehole (Brewster, 1978) proved 13.24 m of grey, generally fissile, calcareous, shelly and pyritic mudstones to a depth of 30.39 m. Seven grey, partially crystalline, cementstones up to 0.10 m thick were penetrated and contain scattered pyrite and shell debris. Secondary selenite is present on some bedding planes. An indeterminate schlotheimiid at 17.34 m may indicate the *angulata* Zone; *Waehneroceras sp.* occurs between 21.14 and 23.37 m and with

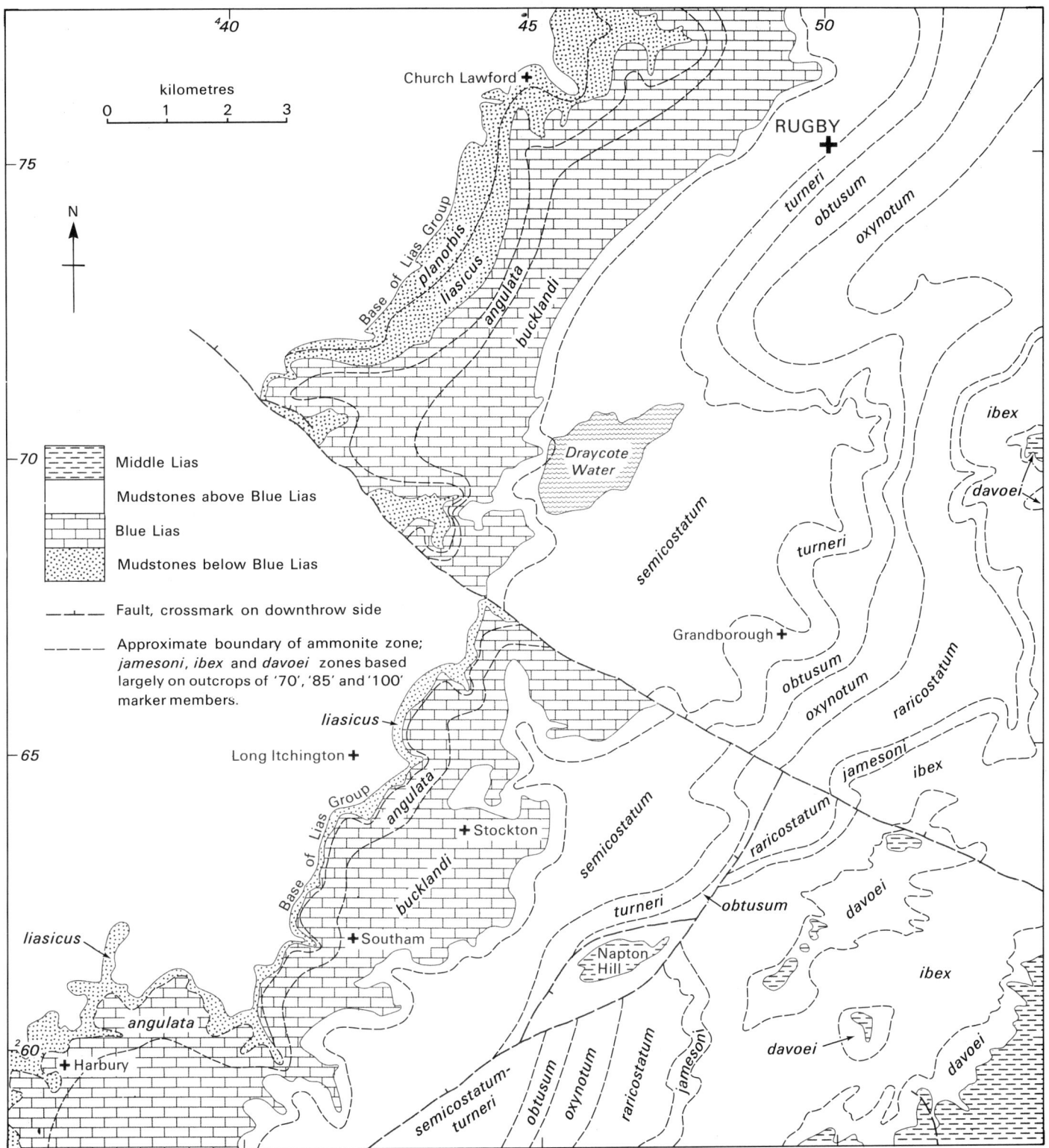

Figure 16   Approximate outcrops of ammonite zones in the Lower Lias

the indeterminate schlotheimiids down to 27.13 m indicates the *liasicus* Zone. The *planorbis* Zone was not proved. Other fossils included *Astarte* cf. *gueuxii*, *Astarte sp.*, *Cardinia sp.*, *Chlamys sp.*, *Liostrea sp.*, *Lucina sp.*, *Modiolus sp.*, *Parallelodon sp.*, *Pseudolimea hettangiensis*, fish fragments and the trace fossils *Chondrites* and *Diplocraterion*.

To the north, in the railway cutting [376 603] at Harbury, the mudstones are obscured by vegetation. Reclassification of sections previously recorded suggests a thickness of 12.3 m (Woodward, 1893; Arkell, 1947) or 17.2 m (Brodie, 1875); the former seems more likely. Brodie (1875) and Woodward (1893) recorded '*Ammonites Johnstoni*', this probably indicates a *Caloceras* from either the *planorbis* Zone (*johnstoni* Subzone) or the *liasicus* Zone.

An old quarry [3926 6096] near Ufton Hill Farm shows the following strata overlying limestone of the Langport Member:

	Thickness m
Mudstone; dark blue-grey, moderately fissile with cementstone nodules up to 0.06 m across near top	1.1
Marl; pale grey, silty, hard	0.1
Cementstone; pale grey, fine-grained, nodular, with hard marl between nodules; passing to	0.1
Mudstone; blue-grey, with silty wisps, pale and marly at top; irregular iron-rich nodules, commonly coated with and containing selenite; scattered small bivalves and ammonites	c.6.0

The ammonites collected were *Waehneroceras?* nuclei from the *liasicus* Zone.

The Long Itchington Quarry B Borehole [4200 6306] and the lowest part of the exposed section together proved a full thickness of 13.17 m of mudstones to a depth of 22.01 m. The Stockton Locks Borehole [4297 6485] (Ambrose, 1978) commenced just below the Blue Lias and proved 10.87 m. Lithologies in the latter are similar to those in Harbury Quarry Borehole; the mudstones are pyritic at their base and contain two cementstone bands, each 0.05 m thick. A specimen of *Waehneroceras sp.* at 5.1 m indicates the *liasicus* Zone, but comparison with Home Farm Borehole suggests that the basal beds could be of *planorbis* Zone age but no faunal evidence for this was found. The bivalve *Lucina sp.* is also present. JB, KA

On the north side of the River Leam north of Birdingbury, debris from a trench showed dark grey, shaly mudstone with fine-grained limestones and cementstones. The ammonite *Waehneroceras sp.* of the *liasicus* Zone was found [4313 6920], with *Schlothemia sp.* of the *angulata* Zone nearby [4318 6920]. KA

Princethorpe to King's Newnham

In Home Farm Borehole resting on the Langport Member at 9.35 m, are 1.25 m of dark grey, fissile mudstone with two thin nodular cementstone beds, a little ramifying pyrite and a few layers of shell debris. The mudstone is overlain by weathered clay and glacial deposits. Fossils included *Liostrea*, *Lucina?*, *Oxytoma?*, *Plagiostoma?*, *Pseudolimea*, crinoid columnals, echinoid spines, foraminifera and ostracods; *Psiloceras planorbis* was abundant between 8.11 and 8.28 m.

'*Ammonites planorbis*' has been recorded from pits at Stretton-on-Dunsmore [?432 732] and Limestone Hall [443 755], the Church Lawford railway cutting [444 759], Long Lawford [462 769] and King's Newnham (Newnham Regis) [464 774]; '*Ammonites johnstoni*' has been recorded from King's Newnham (Cleminshaw, 1868b; Oldham, 1878). Woodward (1893, p.162) recorded 6.1 m of paper shales with pyrite, selenite, small cementstone nodules and echinoid spines overlying the Langport Member in the railway cutting. Oldham (1879, p.46) described the beds at Long Lawford and King's Newnham as coarse, poorly fossiliferous shales containing many layers of selenite crystals. At the King's Newnham pit [4656 7740], 1 m of contorted, dark blue-grey, fissile mudstone with silty lenticles lies beneath terrace gravels. Similar contortion, probably an effect of frost action, was noted by Oldham at Long Lawford.

Thicknesses of mudstones below the Blue Lias in boreholes of the Rugby Portland Cement Co. were as follows: 17.6 m in Lodge Farm No. 1 [4857 7556], 17.8 m in Lodge Farm No. 2 [4859 7580], 18.7 m in Quarry No. 4 [4920 7571], 13.5 m in Quarry No. 5 [4948 7592], 15.6 m in Lodge Farm No. 7 [4828 7556] and 14.2 m in Malpass No. 2 [4879 7631]. In Rugby Waterworks Borehole [5076 7379] (Wilson, 1869; Richardson, 1928, pp.95, 96), these beds were recorded as 11.4 m thick with a 1.2 m limestone near the top. MGS

BLUE LIAS

The Blue Lias consists of alternating dark grey, blocky to fissile, shelly, commonly bioturbated mudstones, and paler grey, hard, argillaceous limestones (cementstones) which are mostly 0.1 to 0.2 m thick and contain organic debris and burrows. Cementstone also occurs as bands of nodules. Within the mudstone units are beds (rarely more than 0.2 m thick) of finely laminated, dark brownish grey paper shale, usually slightly bituminous and commonly associated with pyrite. Calcareous mudstones and marls are particularly common immediately above and below cementstones. Individual cementstones and paper shales show remarkable lateral persistence, and correlations (Figure 17) follow those of Clements (1977). Four cementstone beds provide particularly useful markers: the Top Rock, the Thick Rock, the Worm Bed (after Clements, 1975, 1977), and the Calcirhynchia calcaria Bed (*Rhynchonella* Bed of Clements). The last can be recognised in field brash, and has been mapped locally. The proportion of limestone within the succession decreases northwards from about 35 per cent at Harbury to around 30 per cent at Long Itchington and Rugby.

Approximate thicknesses of the Blue Lias are: 24 m at Harbury, 25 m at Long Itchington, 31 m at Draycote, and 36–40 m in the Rugby area; Napton No. 3 Borehole [4736 6099] proved 35.3 m.

The Blue Lias exhibits a rhythmic alternation of limestone (cementstone) and mudstone or shale, described in detail for the Dorset and Glamorgan coasts by Hallam (1960). The burrowing organisms, in particular *Diplocraterion*, are indicative of shallow water conditions. Precipitation of calcium carbonate increased, and there was some redistribution of lime from calcareous mud to cementstones. The bituminous paper shales, with their improverished faunas, are indicative of an anaerobic environment free of current activity. Hallam's (1967) suggestion that the Lower Toarcian bituminous shales originated in shallow water (possibly less than 20 m) was based on a mathematical model produced by Keulegan and Krumbein (1949). He envisaged an extensive continental shelf of very shallow gradient on which waves dissipated their energy before reaching the shore, thus leaving large areas of sediment subject to little or no wave disturbance. In these areas of minimal water circulation, partial or total stagnation occurred and bituminous muds accumulated. Slight deepening of the water then led to increased circulation, a more aerobic environment and a return to more typical Blue Lias sedimentation.

Cementstones were formed by both primary (sedimentary) and secondary (diagenetic) processes; dark mud-filled burrows extending down into cementstones point to the former, and cementstone nodules passing laterally into marl and mudstone to the latter. Hallam (1964), concluded that the primary lime-rich layers were further enriched by diagenetic migration of $CaCO_3$, citing chemical analyses from the Blue Lias of southern England as evidence. He found (1960, 1964) that the cementstones commonly contain 75–80 per cent $CaCO_3$ and the mudstones 30–56 per cent, with no intermediate values. Analyses by the Rugby Portland Cement Co. show that most samples from the Rugby area conform to these findings, but that a number of cementstones plot out in the 56–70 per cent $CaCO_3$ range.

The age of the Blue Lias in this part of Warwickshire ranges from late in the *Alsatites laqueus* Subzone of the *liasicus* Zone to late in the *Coroniceras rotiforme* Subzone of the *Coroniceras bucklandi* Zone. Its base is diachronous, lying near the top of the *liasicus* Zone at Rugby and younging south-westwards along the outcrop. Apart from in the Rugby area, mudstones below the thickest Blue Lias successions are thinner than elsewhere, possibly indicating earlier onset of Blue Lias facies. It is presumed that older beds are also successively overlapped south-eastwards on to the London Platform. The Hettangian–Sinemurian boundary (base of *bucklandi* Zone) was identified by Clements (1975) in Long Itchington Quarry at the base of his bed 26c. At Rugby over 2 m of sediments separate known *angulata* and *bucklandi* Zone ammonites; Clements (1977) placed the boundary within his bed 27c, by correlation with Long Itchington. No subzones have been delimited in the *angulata* Zone but the presence of *Schlotheimia germanica* in beds up to Clements's 22c at Long Itchington suggests the *Schlotheimia extranodosa* Subzone. The *conybeari–rotiforme* Subzone boundary (*bucklandi* Zone) can be identified at Rugby (Hallam's base of Bed 3 *in* Sylvester-Bradley and Ford, 1968) which is at about the base of Bed 38a of Clements (1977).

The Blue Lias macrofauna is varied, though somewhat sparse at most horizons. Bivalves are most common, and are abundant at some levels. Ammonites, brachiopods and crinoids are usually rare, though locally abundant in some limestones. The following list includes fossils collected during the surveys of the Warwick district and an examination of Harbury Quarry immediately to the south, plus some additional taxa listed by Hallam (*in* Sylvester-Bradley and Ford, 1968) and Clements (1975, 1977), and certain museum specimens: *Calcirhynchia calcaria*, *Spiriferina sp.*, terebratuloids; crinoid columnals; *Anningella faberi*, *Camptonectes sp.*, *Cardinia sp.*, *Chlamys calva?*, *Chlamys sp.*, *Gryphaea arcuata*, *Liostrea hisingeri*, *Lopha sp.*, *Meleagrinella olifex*, *Modiolus sp.*, *Placunopsis striatula*, *Plagiostoma giganteum*, *Pleuromya sp.*, *Plicatula sp.*, *Pseudolimea sp.*, *Pseudopecten priscus*, *Angulaticeras sp.*, *Caloceras bloomfieldense*, *Caloceras sp.*, *Coroniceras* aff. *conybeari*, *C. hyatti*, *C. rotiforme*, *Schlotheimia angulata*, *S. germanica*, *S. sp.*, *Vermiceras scylla*, *V. solaroides* and *Waehneroceras sp.*

The trace fossils *Chondrites* and *Diplocraterion* are common at some levels, and *Rhizocorallium* and *Kulindrichnus* have been found. Also represented are ostracods, plant fragments, nautiloids (probably *Cenoceras*) echinoid plates and spines including *Hemipedina tomesi* from Long Itchington (Woodward, 1893), fragments of fish scales, bones, and reptilian remains. Specimens of the fish *Pholidophorus* and *Dapedium* have been found at Harbury and Stockton respectively; both are in the Warwick Museum collection. Reptilian remains include vertebrae and bones of *Plesiosaurus* and *Ichthyosaurus*. Long Itchington Quarry has yielded a complete skeleton of *Ichthyosaurus* some 3 m in length (Drinkwater, 1912). A complete ichthyosaur skeleton 5.84 m long, was also discovered at Stockton (Spens, 1899) and is illustrated as *Ichthyosaurus* (now *Temnodontosaurus*) *platyodon* in the report of the Rugby School Natural History Society for 1899 (Anon, 1900).

Details

Harbury to Stockton

The Harbury Quarry Borehole [3992 5899] (Brewster, 1978) proved about 14.6 m of Blue Lias to a depth of 17.15 m; commencing just above the Thick Rock, about 10 m below the top of the Blue Lias (Figure 17). Alternating grey, poorly fissile, shelly, bioturbated mudstone, and grey, crystalline cementstone occur together with a few beds of dark grey, bituminous paper shale. The cementstones contain scattered shell debris and burrows of *Diplocraterion* and *Chondrites*, those of the former commonly filled with dark grey mud. A 0.9 m cementstone at 7.81 m correlates with the Worm Bed of Clements (1975; 1977) seen in the Long Itchington and Rugby quarries. The Calcirhynchia calcaria Bed is represented by 0.44 m of cementstone (split by a thin mudstone parting) at 10.37 m; *C. calcaria* is abundant down to 9.96 m, particularly above the mudstone parting. Most of the strata are of *angulata* Zone age; the highest *Schlotheimia sp.* occurring at 7.53 m, but the base of the Zone was not well proved. Correlation with neighbouring sections (Figure 17) suggests that the topmost beds belong to the *conybeari* Subzone of the *bucklandi* Zone. Other fossils found include *C. calcaria* at various horizons down to 10.37 m, also *Anningella*, *Astarte*, *Cardinia*, *Chlamys*, *Liostrea*, *Lucina*, *Modiolus*, *Parallelodon*, *Plagiostoma*, *Pseudolimea* and fish fragments.

In the neighbouring Harbury Quarry 12 m of Blue Lias are overlain by 12 m of mudstone. A composite section is given in Appendix 3. Fossils have been listed by Brodie and Kirshaw (1872), Brodie (1875), Nuttall (1916) and Edmonds and others (1965, p.33). Collecting during the recent survey yielded *Angulaticeras sp.* and *Vermiceras scylla* from the Top Rock. These ammonites, and redetermination of fossils listed by Edmonds and others (1965) indicate the *rotiforme* Subzone of the *bucklandi* Zone, and the suggestion by Edmonds and others that the *resupinatum* [*sauzeanum*] Subzone might be present resting non-sequentially on the *rotiforme* Subzone now seems unlikely. Although the top 10.1 m of the Blue Lias may be referred to the *rotiforme* Subzone, the age of the underlying strata is not proved. Correlation with neighbouring sections (Figure 17) suggests the presence of the *conybeari* Subzone. Warwick Museum has specimens of plesiosaur vertebrae and of *Pholidophorous sp.* from 'Harbury'.

In the Harbury railway cutting only the lowest 9 m of Blue Lias which include the Calcirhynchia calcaria Bed, remain exposed. Brodie's (1875) section, slightly modified, is given in Figure 17. Fossils listed by Brodie (1875) and Woodward (1893, p.160) indicate both the *angulata* Zone and the *conybeari* Subzone of the *bucklandi* Zone.

Beds near the top of the Blue Lias dip 4° SSW in the banks of the River Itchen [415 588] at Ladbroke:

	Thickness m
Cementstone; pale grey to cream, weathered and fractured	0.03–0.07
Clay; brown-grey	0.23
Cementstone; pale grey, weathered and fractured into blocks	0.05–0.11
Clay; grey-brown	0.03–0.10
Cementstone; grey, fine-grained, with rare finely comminuted shell debris	0.13
Clay; pale brown-grey, with rare concretions up to 20 mm across and ferruginous patches	0.55
Cementstone; pale to medium grey, with irregular weathered marly top and dark staining along some joints	0.20
Clay; pale grey-brown	0.26
Cementstone; pale grey, weathering to orange, with a little shell detritus; ripple mark traces	c.0.2

Figure 17 The correlation of the
Blue Lias

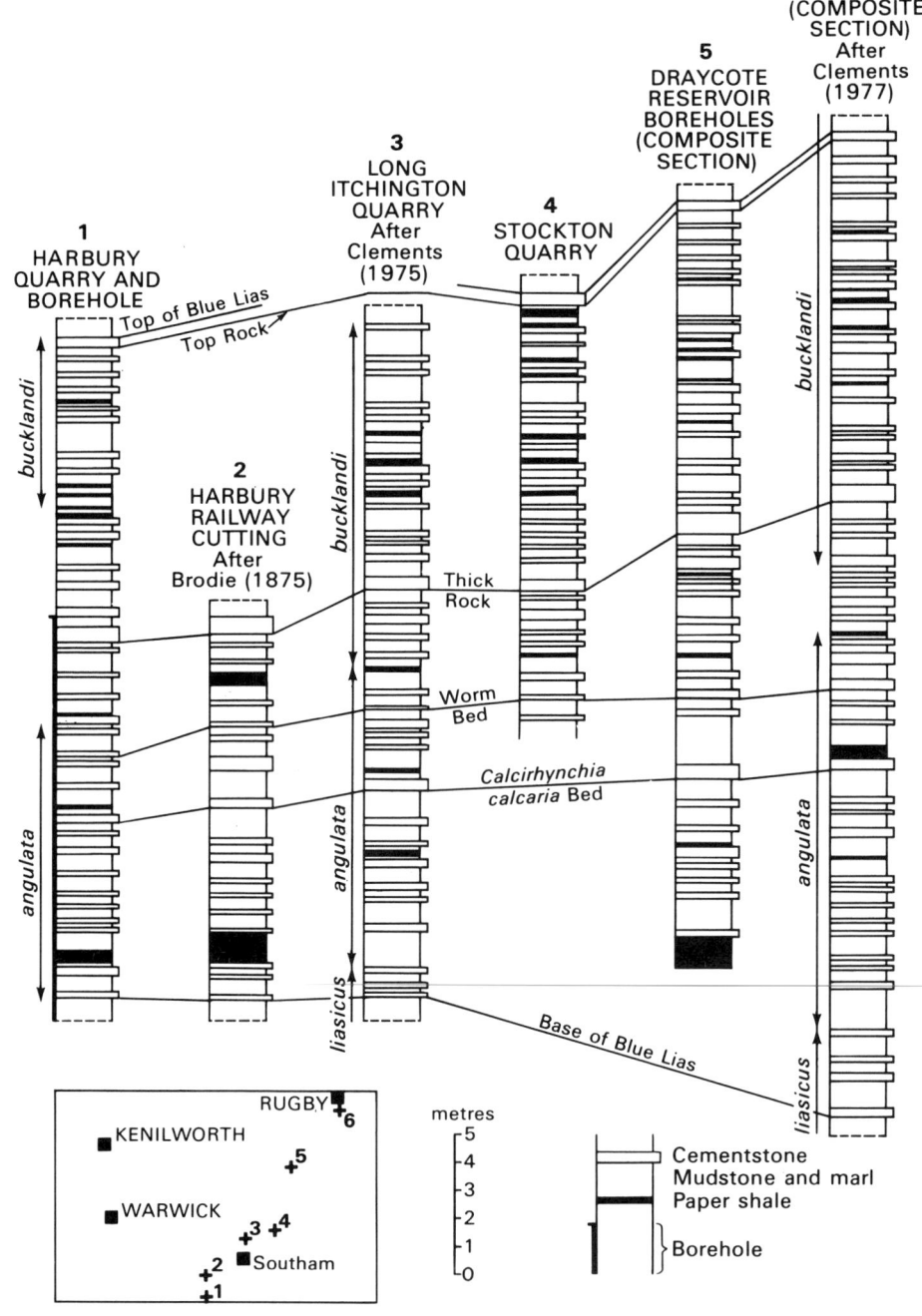

JB, KA

Long Itchington Quarry [around 41 to 42 63] provides a sequence of the Blue Lias up to just below the Top Rock (Figure 17). The quarry section (Clements, 1975) shows the *Rhynchonella* Bed (now termed the Calcirhynchia calcaria Bed), the Worm Bed and the Thick Rock.

Clements found *Waehneroceras sp.* and *Caloceras* cf. *bloomfieldense* in debris from the base of the quarry face, and Tutcher (*in* Buckman, 1918) reported *Waehneroceras extracostatum* from the base of the face. Thus the *laqueus* Subzone of the *liasicus* Zone is present. The basal 11 m of the face yielded many specimens of *Schlothemia sp.* from between beds 26b and 14a and are assigned to the *angulata* Zone. The mudstone bed 26c of Clements yielded *Vermiceras* cf. *solaroides* indicating the *conybeari* Subzone of the *bucklandi* Zone. *Coroniceras rotiforme* was found about 1.7 m from the top (bed 40g) and with *C. hyatti* in bed 38a (of Clements) indicate the *rotiforme* Subzone. Other

fossils include *Calcirhynchia calcaria, Spiriferina, Gryphaea arcuata, Plagiostoma giganteum, Pleuromya* and fragments of echinoids, crinoids, fish and rare reptilian remains. Abundant foraminifera and ostracods occur. Trace fossils including *Diplocraterion* and *Chondrites* are abundant in some beds, and *Kulindrichnus*, believed to be the trace of a burrow or resting trail of a cerianthid sea anemone (Hallam, 1960), occurs in limestones.

A quarry [440 640] at Stockton shows a complete sequence from just below the Worm Bed to the Top Rock (Appendix 3); correlation with nearby sections indicates a range from late *angulata* to late *bucklandi* Zone. Another [438 649] shows beds below the Top Rock lying mainly within the *rotiforme* Subzone but possibly extending into the underlying *conybeari* Subzone (Appendix 3). Nuttall's (1916) ammonites from Stockton suggest the presence of both the *angulata* and *bucklandi* zones.

Stockton to Birdingbury

The Calcirhynchia calcaria Bed is persistant between Stockton and Leamington Hastings, together with an overlying cementstone rich in *Gryphaea arcuata*. *Plagiostoma gigantea* and *Pentacrinites* occur in lenticular pockets in one or more cementstones above the Calcirhynchia calcaria Bed. Ditch sections have yielded *Schlotheimia sp.* (*angulata* Zone) [4404 6679] from about 4 m above the base of the Blue Lias, and *Waehneroceras sp.* (*liasicus* Zone) [4348 6622] from just above the base of the Blue Lias.

The upper of two cementstones exposed in another ditch [456 663] contains *Calcirhynchia calcaria*, crinoids, and *Coroniceras?*, probably from the *rotiforme* Subzone of the *bucklandi* Zone. An anomalous south-west dip is attributable to the effects of the Princethorpe Fault, near which the cementstones commonly contain much ramifying calcite.

The Birdingbury Hall Borehole [4322 6869] proved Blue Lias to 14.7 m, strata erroneously classified by Richardson (1928, p.49) as Lower Lias, Rhaetic and Keuper Marl. To the east, a ditch [4447 6846] contains 6 m of alternating thin clays and cementstones; *Gryphaea arcuata* occurs at several horizons, *Pentacrinites* in the Calcirhynchia calcaria Bed, and *Coroniceras?* just above it. Farther north, in a railway cutting [4361 6965], 2.5 m of alternating thin cementstones and mudstones lie near the base of the Blue Lias; they yielded an indeterminate coroniceratid ammonite and the trace fossil *Chondrites*.   KA

Draycote to Rugby

Boreholes drilled on the site of Draycote Reservoir [460 700] provide a composite section (Figure 17) of almost the whole of the Blue Lias which is probably about 31 m thick. *Coroniceras rotiforme* in Draycote No. 4 Borehole [4519 7003], *C.* cf. *rotiforme* in Draycote No. 21 [4607 7010] and *Coroniceras sp.* in Draycote No. 23 [4591 6960] occurred at levels corresponding with the *C. rotiforme* Subzone in Rugby Quarry.

The Blue Lias of Rugby Quarry [493 759] has been described by Clements (1977), and the upper beds by Hallam (*in* Sylvester-Bradley and Ford, 1968, p.206). Virtually a full succession (of 36 m) is exposed (Figure 17; Plate 6). In the north-western part of the quarry the beds are disposed in a number of anticlinal flexures, commonly fractured at the crests; these extend only a few metres laterally, and die out downwards within 10 m of the surface. Similar structures in New Bilton Claypit [?492 756] (Wilson, 1870b, p.196) affected sediments of the Fourth Terrace; they are probably a result of frost action during the Pleistocene. The Blue Lias formerly exposed in the Victoria Quarry, now the site of the cement works [487 756], has been described by Lowe (1873), Oldham (1878) and Woodward (1893), and reported as yielding *Ichthyosaurus* and *Plesiosaurus*.

The top 2.5 m of the Blue Lias are exposed above water level in Newbold Quarry [494 770] and resemble the corresponding section in Rugby Quarry. Woodward (1893) described 13.7 m of blue shales and grey argillaceous limestones, more clayey and browner in colour towards the top. He figured an anticline which, although larger than those of Rugby Quarry, is probably another superficial structure. Geological Survey field maps made in 1915 record 21 m of grey, shaly clay alternating with cementstones. Oldham (1878) listed '*Ammonites angulatus*, *Ichthyosaurus* and *Teleosaurus*' from Newbold Quarry, indicating the presence of the *angulata* Zone. The 1915 field maps record quarry sections showing 4.6 m [478 771] and 6.6 m [484 769] of grey clay with cementstone. Oldham's (1878) record of '*Ammonites angulatus*' from Newbold Old Works probably refers to the former locality.

The Blue Lias of the Rugby Waterworks Borehole (Wilson, 1869; Richardson, 1928, pp.95, 96) was 45 m thick; the lowest 18 m match the section in Rugby Quarry, with equivalents of the Calcirhynchia calcaria Bed and the Worm Bed.   MGS

MUDSTONES ABOVE THE BLUE LIAS

The strata overlying the Blue Lias comprise 150 to 170 m of dark grey, blocky to fissile, calcareous, fossiliferous, commonly pyritic mudstones with a few cementstone beds and cementstone and ironstone nodules. In the upper part of the sequence, beds of grey, shelly, detrital limestone occur with calcareous mudstones and marls. Individual limestones are usually impersistent, but two limestone – marl intercalations persist laterally and have been correlated with the '70' Marker and '85' Marker Members of Horton and Poole (1977). Both were proved in the Barby Borehole, immediately east of the district (Ambrose and Brewster, 1979; Ambrose and Ivimey-Cook, 1982). A third, the '100' Marker Member, may be present locally in the Flecknoe – Upper Shuckburgh area where several limestones occur above the '85' Marker.

From the '70' Marker upwards to the Middle Lias the mudstones become more silty and micaceous and paler grey in colour. Silty strata adjoin the '85' Marker west and south of Barby Hill. Towards the top of the Lower Lias these beds are ferruginous in parts, with more abundant ironstone nodules. As a consequence during the survey of the Northampton district (Sheet 185) the top of the Lower Lias was mistakenly taken at the '85' Marker.

The mudstones above the Blue Lias contain a rich and varied fauna. Full lists are held in BGS files, but see also details below.

The *semicostatum* Zone of the Warwick district may be 30 m thick, so is amongst the thickest known in the Midlands. The succeeding *Caenisites turneri* and *Asteroceras obtusum* zones are poorly known and the thicknesses given in Table 3 are tentative. Non-sequences occurred within the *raricostatum* and *jamesoni* zones in Barby Borehole; the *jamesoni* Zone was markedly attenuated and lay wholly within the '70' Marker. The *Tragophylloceras ibex* Zone thins southwards from Barby. The *Prodactylioceras davoei* Zone thickens westwards from about 18.6 m at Barby, to at least 25 m at Napton on the Hill (Callomon *in* Sylvester-Bradley and Ford, 1968, p.205).

Palaeontological evidence at outcrop is generally sparse. Many Lower Lias fossils from the district are housed in Warwick and Northampton county museums. Lists have been given by Woodward (1893, 1897), Thompson (1899), Nuttall (1916) and Hallam *in* Sylvester-Bradley and Ford (1968). The fossils are mostly bivalves, together with brachiopods, ammonites, belemnites, gastropods, echinoid fragments, fish and trace fossils.

The mudstones above the Blue Lias originated in deeper water than the preceding strata. They resemble the earliest Liassic mudstones but contain a richer fauna. Small-scale transgressive cycles are characteristic; each begins at a non-sequence and comprises calcareous mudstones whose calcium carbonate content decreases upwards.

The '70' Marker indicates a significant, though short-lived, environmental change. Shallow waters were kept well aerated by circulating currents. Shells and shell debris form much of the sediment, pointing to increased precipitation of calcium carbonate and very little influx of terrigenous mud. This marker extends over wide areas of the south Midlands (Horton and Poole, 1977).

Plate 6   Interbedded limestone and mudstone of the Blue Lias, passing upwards (top centre and right) into mudstones. The face is 45 m high. Rugby Quarry.   A13079

The succeeding strata represent a return to mudstone sedimentation, but the incoming of increasing amounts of coarser terrigenous sediment, in the form of silt and mica, herald localised current-activated shallow water which produced the shelly detrital limestones and cementstones of the '85' Marker. Mudstones at the top of the Lower Lias were laid down in shallow water near to land. They are characterised by increased silt content, local iron precipitation, burrowing and periods of current activity. There are some shell pavements and impersistent beds of shell detrital limestone. The conditions responsible for the '100' Marker, which were widespread farther south, did not extend throughout the present district, however a minor nonsequence found in Barby Borehole at 29.8 m (Ambrose and Ivimey-Cook, 1982) may represent this Marker.

Details

South-west of Princethorpe Fault

Zonal evidence from strata below the '70' Marker was obtained from small exposures as follows: *Arnioceras sp.* and *Euagassiceras resupinatum* [4783 6363; 4758 6369], indicating the *resupinatum* Subzone of the *semicostatum* Zone; *Gryphaea?*, *Hippopodium ponderosum* and *Caenisites?* [4515 6150], probably from the *turneri* Zone; *Arnioceras sp.* [4255 5931; 4268 5924] and *Coroniceras sp.* [4762 6376], indicating a

high *bucklandi* to *obtusum* zonal position.

The lowermost 12 m of these beds exposed at Harbury Quarry, are described in Appendix 3 (p.81).

Complete, but uncored, sequences through these lowermost beds were penetrated by Napton No. 4 and No. 5 and the In Meadow Gate [4850 6036] boreholes; geophysical logs indicate thicknesses of 86.9 m, 89.9 m and about 87 m respectively, below the '70' Marker. These beds probably range from late *bucklandi* Zone to *jamesoni* Zone in age, and appear to include several limestones or calcareous mudstones in their lowest parts. The '70' Marker was about 3.1 m and 3.7 m thick respectively in Napton No. 4 and No. 5 boreholes. To the south-east it forms a sharp feature, but was exposed only in a temporary trench [4725 5874] where pale grey silty clay contained fragments of shell detrital limestone and cementstone with belemnites, echinoid spines and *Gryphaea sp.* This limestone feature was interpreted as the wave-cut bench of a Pleistocene lake by Dury (1951).

The immediately overlying beds are poorly exposed, though Napton No. 3, No. 4 and No. 5 boreholes proved that the strata between the '70' and '85' Markers are of similar thickness (35 m) to those in the Barby Borehole and belong to the *valdani* Subzone of the *ibex* Zone. Sparse evidence shows the '70' Marker to be overlain by silty, micaceous mudstones between Marston Doles and Flecknoe. Sections published by Woodward (1897) and Thompson (1899) of cuttings of the old Great Central Railway were interpreted by Green and Melville (1956) as indicating a thick sequence of *jamesoni* Zone strata overlain by a thin sequence of *ibex* Zone age.

Limestone features about 30 m below the Middle Lias and 30–40 m above the '70' Marker are thought to include the '85' Marker. Napton No. 4 and No. 5 boreholes both commenced in the '85' Marker, respectively 35 m and 39 m above the '70' Marker. The limestone thought to be the '85' Marker caps a scarp running northwards from Marston Doles. The form of the scarp features suggests a gentle south-easterly dip south-west of the Napton – Northfields Farm – Priors Marston road, and a gentle north-westerly dip north-east of it. Surface brash is of grey-weathering blue-hearted, ferruginous, shelly detrital limestone. JB

Above the '85' Marker Green and Melville (1956) interpreted Thompson's (1899) records of the cutting of the old Great Central Railway to suggest a thickness of about 30 m of *davoei* Zone strata. The old brickpit [457 612] at Napton Hill exposed the top of the Lower Lias. Callomon (*in* Sylvester-Bradley and Ford, 1968, pp.205–206) assigned all 25.8 m of the blue-grey clays with ferruginous nodules to the *davoei* Zone, with all three subzones being recognised. The lowest 12.9 m of beds were assigned to the *maculatum* Subzone, but contained diagnostic ammonites only in their uppermost 0.4 m. Comparison with the Barby Borehole succession suggests that the *ibex* Zone may be represented in the pit. The following fauna was collected by R. V. Melville and F. G. Dimes in 1957: *Furcirhynchia* sp., *Rudirhynchia rudis*, *Tetrarhynchia?*; pentacrinoid columnals; *Arcomya* sp., *Camptonectes mundus*, *Goniomya hybrida*, *Grammatodon insons*, *Gryphaea gigantea*, *Liostrea* sp., *Modiolus scalprum*, *Nuculana (Rollieria) bronni*, *Nuculana (Ryderia) doris*, *N. (R.) graphica*, *Oxytoma inequivalvis*, *Palaeoneilo galatea*, *Parainoceramus* sp., *Parallelodon buckmani*, *Pleuromya costata*, *Pronoella* sp., *Protocardia truncata*, *Pseudolimea* sp., *Pteromya* sp., *Tutcheria* sp.; *Amberleya (Eucyclus) imbricatus*, *Zygopleura* sp.; *Aegoceras (Oistoceras) angulatum*, *A. (O.)* cf. *figulinum*, *A. (O.) wrighti*, *Aegoceras* sp., *Androgynoceras* cf. *artigyrus*, *A.* cf. *brevilobatum* and *A. capricornus*.

The '100' Marker Member, may be present north-west of Hellidon, and between Upper Shuckborough and Flecknoe, where several limestone features have been mapped. JB

North-east of the Princethorpe Fault

Field brash [449 674] from near the base of the sequence south-east of Leamington Hastings yielded the bivalves *Astarte*, *Gryphaea*, *Plicatula*, *Pseudolimea*, *Myoconcha decorata*, and indeterminate pectinids, together with *Spiriferina walcotti*, gastropods and ostracods. The ammonites *Arnioceras* sp. and an indeterminate arietid indicate a position between late *bucklandi* Zone and *obtusum* Zone, probably the *semicostatum* Zone.

Of several boreholes sunk during the construction of Draycote Reservoir. No. 7 [4508 6941] proved 27.1 m of strata above the Blue Lias; the lowest 9 m were cored, and comprised interbedded fissile and blocky mudstones with a few thin cementstones. Ammonites indicative of the *semicostatum* Zone or early *turneri* Zone were: *Arnioceras* cf. *bodleyi* between 18.9 and 21.4 m and *Angulaticeras* at 25 m in No. 7 borehole; *A. bodleyi* between 14.6 and 14.9 m in No. 8 [4511 6912] and *A. bodleyi* at 17.7 m in No. 1 [4523 7028] and between 24.4 and 24.5 m in No. 12 [4587 6903]. Surface brash on the southern shores of the reservoir [452 690; 455 690 and 458 691] yielded a *resupinatum* Subzone fauna comprising *Angulaticeras* sp., *Arnioceras* sp., *Euagassiceras resupinatum*, *Cenoceras* sp., *Grammatodon* sp., *Gryphaea arcuata*, *Oxytoma inequivalvis* and belemnites. Additional fossils collected during construction were *Lucina* sp., *Pseudolimea* sp., *Coroniceras?*, *Antiquilima antiquata*, *Cardinia hydrida*, *Gryphaea* cf. *maccullochii*, *Cymbites laevigatus*, pentacrinoid columnals and echinoid fragments; they indicate the *bucklandi* or *semicostatum* zones. Specimens of *E. resupinatum* and *E. subtaurus* found at about 91 m above OD on Hensborough Hill [460 690] indicate the *resupinatum* Subzone.

Exposures of the *resupinatum* Subzone between Leamington Hastings and Grandborough yielded *E. resupinatum* and *Arnioceras* sp., mostly as debris from streams and ditches. About 1.5 m of dark blue-grey, blocky to fissile mudstone with ironstone and limestone nodules, exposed beneath alluvium in the banks of Millholme Brook [460 670], lie near the junction between the *scipionianum* and *resupinatum* subzones. The fauna included *Agassiceras* sp., (aberrant form), *Arnioceras semicostatum*, *A.* sp., *Euagassiceras resupinatum*, *E. subtaurus* and the bivalves 'Gervillia', *Gryphaea* and *Plagiostoma*. Several fossil localities in the *resupinatum* Subzone [497 696 to 498 695; 4975 6975; and 499 697] all yielded *Arnioceras* sp., *E. resupinatum*, and *Gryphaea* sp.; the last locality also yielded *A.* cf. *semicostatum*, *Euagassiceras* sp., *Piarorhynchia*, cf. *radstockiensis*.

A cutting [521 692 to 523 678] for the former Great Central Railway north of Willoughby, now largely overgrown, was described by Woodward (1897) and Thompson (1899) and the beds assigned to the '*armatus*' and '*jamesoni*' zones. The '*armatus*' Zone, (of Oppel) was of early *jamesoni* Zone age (Dean and others, 1961). The strata described were clays with ferruginous layers in their lower part and containing many argillo-calcareous nodules of highly irregular shapes together with a few concretionary ironstone nodules of more regular form. The Oxford Canal runs close by the railway here, also in a cutting, and landslip has exposed dark brown to grey, slightly silty clay with *Gryphaea* sp. A limestone feature running into the cutting is correlated with the '70' Marker and it is likely that the *raricostatum* Zone is represented in the underlying beds. KA

The M45 motorway cutting at Dunchurch was described by Callomon (*in* Sylvester-Bradley and Ford, 1968, p.194) as containing beds of the *Arnioceras semicostatum* Zone in its western half [478 709] and beds of the succeeding *Caenisites turneri* Zone in its eastern half [486 707]. Fossils collected during construction of the motorway indicated the *resupinatum* Subzone of the *semicostatum* Zone [4810 7089]; the *C. turneri* Zone [4823 7087]; the *Caenisites brooki* and *Microderoceras birchi* subzones of the *turneri* Zone [487 708]; probable *turneri* Zone at [4974 7068]; the *obtusum* Zone [5010 7077]; the *resupinatum* Subzone [5049 7079] and a horizon in the *semicostatum – obtusum* zonal range at [5106 7108].

In Rugby Waterworks Borehole (Wilson, 1869; Richardson, 1928, pp.95, 96), about 55 m of blue clay with a few limestone bands overlay the Blue Lias. A limestone at 40.5 m has not been found elsewhere, but a 0.25 m limestone at 17.3 m (i.e. 100.9 m above the base of the Lower Lias) may correspond to a bed exposed in the Rains Brook valley 2 km to the south [c.508 715].

The south-east face of Rugby Quarry shows the basal 8 m of grey shaly mudstone, with thin or discontinuous cementstone beds at 1 m, 3 m and 5 m above the base. Hallam (*in* Sylvester-Bradley and Ford, 1968, p.206, beds 14–21) assigned these beds to the *bucklandi* Zone; the redefinition of this Zone (Ivimey-Cook and Donovan, in Whittaker and Green, 1983, pp.129, 130) assigns these beds to the *lyra* Subzone of the *semicostatum* Zone and the top (Bed 14) of the *rotiforme* Subzone.

Oldham (1879, p.45) described claypits at New Bilton, Bromwich's Pit [492 756], ?Parnell's Pit [499 757] and Pinfold's Pit [495 757], as showing firm dark blue clay abounding in '*Ammonites semicostatus* and *Pleurotomaria anglica*'. Dunchurch Road claypit [502 744] and its fossils have been described by Cleminshaw (1868a, p.35, 1868b), Oldham (1878, 1879, p.45) and Wilson (1870a, p.26). A temporary excavation [5007 7461] just north-west of the site of the pit showed 1 m of grey clay with small ironstone and cementstone nodules and a few fossils indicating a horizon within the *semicostatum – obtusum* zonal range.

Opposite the entrance to Rugby railway station [5105 7580] 3 m of blue-grey shale with cementstone nodules are exposed. To the north, in the bank of the River Avon [c.510 765], Nuttall (1916) recorded 3.7 m of dark blue clay with scattered limestone nodules and selenite and fossils indicative of the *semicostatum* Zone. Wilson (1874) recorded '*Ammonites brookii*' and '*A. birchii*' from a railway cutting [5155 7577 to 5196 7557] at Rugby; this suggests the *turneri*

Zone. The southern part of the Great Central Railway cutting through Rugby [5145 7408 to 5164 7277] has yielded fossils within the *obtusum – jamesoni* zonal range (Fox Strangways, 1897, pp.58, 59). Thompson (1899) recorded the section and described beds of the '*armatus*' Zone (i.e. of early *jamesoni* zone age), which crop out mainly south of the B4429 road bridge, as clays with nodules and much pyrite. The old Upper Hillmorton [5265 7375] and Lower Hillmorton [5332 7389] claypits exposed clays and micaceous shales with calcareous and ferruginous nodules yielding fossils within the *obtusum – jamesoni* zonal range (Cleminshaw, 1868a; Lawe, 1869; Oldham, 1879; Woodward, 1893).

An exposure in the bank of the Rains Brook [5208 7213] shows 2 m of grey shale with rare cementstone nodules. A prominent feature on the valley sides [527 724] probably marks a limestone bed about 15 – 20 m above that noted farther west in the Rains Brook valley [508 715] and in the Rugby Waterworks Borehole. MGS

North-west of Flecknoe, limestone debris yielded *Platypleuroceras sp.*, indicating the middle part of the *jamesoni* Zone, together with *Gryphaea maccullochii arcuatiforme* and *G. maccullochii*; this points to a likely correlation with the '70' Marker. JB

Limestone debris [5284 6472] from south of Wolfhampcote contains *Platypleuroceras sp.*, *Gryphaea* and belemnites, an assemblage typical of the '70' Marker. Dark grey locally ferruginous mudstone exposed in the south bank of the Oxford Canal [523 653] contains a rubbly calcareous layer with limestone nodules; *Gryphaea sp.*, pectinids and belemnites were again found. Arkell (1933, p.134) recorded the coral *Montlivaltia mucronata* from the *jamesoni* Zone near Wolfhampcote, presumably in or near the '70' Marker; which forms a sharp feature at Onley Fields Farm [518 697] along which hard, dark blue-grey, shell fragmental limestone and cementstone with abundant *Gryphaea sp.* crops out. KA

A few metres above the '70' Marker west of Barby Hill dark grey smooth clays are overlain by paler grey slightly silty and micaceous clays. This colour change corresponds to one at 76.07 m depth in the Barby Borehole but is of limited stratigraphical significance. The beds between the '70' and '85' Marker members thicken northwards from north-east of Wolfhampcote (15 – 18 m) to west of Barby Hill (about 25 m) with a further eastwood thickening to 35.43 m in Barby Borehole. Grey, silty, micaceous clays with calcareous and ferruginous nodules in an old brickpit [540 659] just east of the district yielded ammonites (Woodward, 1893, p.166) now redetermined as *Liparoceras cheltiense* and *Acanthopleuroceras valdani*, a fauna indicative of the *valdani* Subzone of the *ibex* Zone. KA

The unit mapped as limestone south of and around Barby Hill comprises about 8 m of silts, silty clays and thin discontinuous limestones and is commonly marked by springs at its base. It probably includes the 0.11 m limestone proved 3.68 m above the '85' Marker Member in Barby Borehole, and also some beds below the Member. The surface brash is predominantly of blue-grey, shell detrital limestone, with a little micritic limestone and cementstone.

The following section was measured in a ditch [5335 6852 – 5335 6857] SSE of Willoughby Lodge:

	Thickness m
Clay; grey, silty	0.3
Silt; brown and orange-stained, ferruginous, muddy, with nodular ferruginous shelly limestones; bivalves and belemnites	0.75
Clay; pale grey to brown, ferruginous, very silty, micaceous	0.25 – 0.3
Limestone; dark blue-grey with brown iron staining; shells and shell detritus; sparry cement	up to 0.1
Clay; pale grey to brown, ferruginous, very silty, micaceous	0.4

Limestone; brown to grey; shells and shell detritus; sparry and micritic cement; crinoid-rich in parts; possible lateral passage to	0 – 0.4
Cementstone; medium grey, shelly, nodular, with bivalves, belemnites and ammonites	0 – 0.1
Clay; pale grey-brown, ferruginous, silty, micaceous; nodular cementstone 0.3 m from top	0.8

Fossils included crinoid columnals, *Piarorhynchia?*, *Parallelodon sp.*, *Pleuromya costata*, *Pseudopecten*, *Liparoceras* cf. *bronni* and a belemnite, indicating the *ibex* Zone. A small exposure [5333 6957] near the base of the mapped unit yielded crinoid fragments, *Gibbirhynchia* cf. *muirwoodae*, *Cardinia sp.*, *Entolium liasianum*, *Gryphaea sp.*, *Plicatula spinosa*, *Pseudopecten?*, *Amberleya* and belemnite fragments. KA

North of Barby Hill, the '85' Marker is represented by two thin limestone beds about 2 m apart, which form small features; the lower consists of nodular ferruginous cementstone, and the upper of yellow-weathering, blue-grey, shelly limestone. MGS

Above the '75' Marker Thompson's (1899) record of the '*Henleyi*' (*davoei*) Zone in the railway cutting east of Flecknoe was confirmed by the finding [5233 6323] of *Androgynoceras* cf. *maculatum* (*maculatum* Subzone), together with *Cuneirhynchia oxynoti*, *Astarte sp.*, *Modiolus scalprum* and *Aegoceras sp.* in blue-grey, slightly silty, micaceous clay with an associated thin, impersistent, shelly detrital limestone and scattered ironstone nodules. Another exposure [5241 6347] shows:

	Thickness m
Clay; blue-grey, slightly silty and micaceous	0.5
Ironstone nodule	0.04
Clay; grey, slightly silty with ferruginous lenticles and staining; 0.05 m of shelly detritus at 0.6 m underlain by impersistent ironstone	about 1.0
Limestone; shell detrital	0.04
Clay; blue-grey, slightly silty	0.5

Thompson (1899) recorded '*Ammonites ibex*' and '*A. valdani*' from the base of the cutting [524 641] east of Nethercote, and placed the top beds in the '*Henleyi*' (*davoei*) Zone. Reference to the position of the *davoei* Zone (Figure 16) and to the occurrence of *Platypleuroceras sp.* of the underlying *jamesoni* Zone [5284 6472] (p.43) implies a thickness of only 15 m for the *ibex* Zone hereabouts.

Field evidence around Barby Hill shows that the mudstones overlying the '85' Marker become more silty and iron-stained upwards, with more ironstone nodules. The nodules, commonly orange-brown, concentrically ringed, and limonitic, are probably altered from siderite-calcite cementstones. Some are sandy and exhibit box-weathering, with thin films of orange limonite. KA

Barby Borehole

Some 37.7 m of dark grey, fissile, shelly mudstones with burrows, pyritic trails, disseminated pyrite and cementstone nodules, underlying the '70' Marker, were proved in the Barby Borehole; they include beds belonging to the *oxynotum* Zone (Ambrose and Ivimey-Cook, 1982). Sedimentary rhythms are present; each comprises calcareous, blocky mudstone, with abundant shell material commonly as pavements, a few phosphatic pebbles, and scattered *Chondrites*, passing up into smooth, less shelly mudstone cut off by a non-sequence associated with abundant *Chondrites* and other burrows.

The '70' Marker is 4.46 m thick at 82.31 m depth and comprises grey, fossiliferous marl, with scattered nodules of cementstone and shell detrital limestone and layers of calcareous mudstone. Non-sequences lie at both top and bottom, and others are present within the Member. Shells, commonly pyritised, comprised mostly bivalves, together with a few ammonites, belemnites, gastropods, brachiopods and crinoids. Burrows are common. The Member includes the whole of the *jamesoni* Zone.

The Member is overlain at 77.85 by 35.43 m of smooth, micaceous, shelly, fissile mudstones, commonly burrowed with a few intensely bioturbated horizons, containing scattered nodules of cementstone and pyrite. The upper beds are slightly paler grey than the lower; they also contained more nodules and a little silt. Several cementstone nodules have dark centres; two, from depths of 47.11 and 48.23 m, were analysed by Mr B. R. Young who reported that the dark material was a mixture of calcite and carbonate apatite (X-ray powder films X 8028–8029). The lowest 1.78 m of the beds are calcareous, and the bottom 0.2 m contain abundant shells, many pyritised. The mudstones above 50.7 m are calcareous and micaceous.

The '85' Marker, 1.77 m thick, is proved at 42.42 to 40.65 m. It consists of three shell detrital limestones interbedded with pale grey, silty, blocky, bioturbated, calcareous mudstones containing abundant shells and shell detritus, a few cementstone nodules, pyritic trails, and large burrows filled with shell detritus. Ill-defined sedimentary rhythms show an upward decrease in carbonate content. The base of the Member is taken at the bottom of the lowest limestone and the top at the top of the highest. It is overlain by blocky to poorly fissile, silty mudstone with less silty layers, a few thin limestones and cementstones, and scattered nodules of cementstone, siderite and pyrite. A 0.11 m hard, grey, shelly, argillaceous limestone occurs at 36.74 m. Shells and shell detritus are common; some form pavements representing minor non-sequences, and other non-sequences occurred at 35.94 and 29.80 m. Burrows are commoner and silt content greater above 31.5 m. Pyritic trails are found throughout. Layers of brecciated mudstone occur above 28.3 m, and are dominant above 19.5 m; possibly they reflect intense bioturbation. *Chondrites* and similar worm mottlings are common. The larger horizontal burrows are usually filled with pyritised silt, dark grey mud, or shell detritus. Shells and shell detritus are less common above 19.33 m, and organic remains rare above 15 m except for a few plant and wood fragments. Bivalves are the most abundant fossils, with ammonites common only locally; crinoids, gastropods, belemnites, brachiopods and fish fragments are also found. The '100' Marker may be represented by a non-sequence at 29.8 m.   KA

MIDDLE LIAS

The Middle Lias crops out mainly in the south-eastern extremity of the district where the Marlstone Rock Bed forms the crest of a major escarpment; outliers occur at Napton on the Hill, Upper Shuckburgh and Flecknoe. The beds below the Marlstone Rock Bed comprise mostly siltstones and silty mudstones with impersistent beds of ferruginous limestone, some of them oolitic, and sandstone. Their junction with the underlying Lower Lias is generally transitional and marked by springs. The Marlstone Rock Bed consists of a ferruginous, sandy, shell detrital limestone which is oolitic in patches. Ooliths are less common than in the main ironstone field to the south, but some fresh surfaces show the dark blue-green colour associated with the chamosite ooliths of the Banbury district.

The siltstones and silty mudstones are 23.3 m thick at Napton (Howarth, 1958) and 25–30 m on the main escarpment. A gamma-ray log of the Barby Borehole suggests about 9.6 m of Middle Lias strata; the borehole is shown on the published geological map (Sheet 185, Northampton), as having started in Marlstone Rock Bed but this is uncertain. Thompson's (1889, pp.6–7) generalised section for North-

amptonshire showed 14–15.5 m of Middle Lias including 1.8 m of Marlstone Rock Bed. The Marlstone Rock Bed of the Warwick district ranges from 2.5 to 4 m thick. In general, the Middle Lias is thinner than in neighbouring areas, and evidence points to a condensed sequence rather than to erosion or lateral passage (Hallam *in* Sylvester-Bradley and Ford, 1968).

Palaeontologically the siltstones and silty mudstones correspond to the *Amaltheus margaritatus* Zone, proved only in the old Napton Claypit (p.46). Because the lithological junction with the Lower Lias is transitional the mapped base probably falls within the underlying *davoei* Zone. The Marlstone Rock Bed spans the Pliensbachian–Toarcian boundary. Its lower part corresponds to the *Pleuroceras spinatum* Zone, and Howarth (1978, 1980) has demonstrated that its upper part spans the three lowest subzones of the *Dactylioceras tenuicostatum* Zone. Middle Lias fossils collected during the recent survey are mostly bivalves together with ammonites and crinoid debris. Other specimens are housed in the Warwick Museum.

Middle Lias times witnessed the conclusion of a major regressive cycle, with deposition of clays, silts and fine sands in a shallow sea favourable to the precipitation of iron as siderite and chamosite. The presence of ooliths in the Marlstone Rock Bed, and of a conglomerate locally at its base, are indicative of strong currents and a period of erosion. The source of iron is presumed to be a nearby well-vegetated land mass experiencing a warm humid climate; iron was concentrated by intense leaching, and carried into the sea (Hallam, 1975).

Details

Napton on the Hill to Flecknoe

The old brickworks [4577 6129], in the outlier at Napton Hill formerly provided a complete section of the Middle Lias and the *davoei* Zone of the Lower Lias. The only published section is that of Howarth (1958, p.xi), quoted by Sylvester-Bradley and Ford (1968, p. 205). Only the top 20 m of the Middle Lias were visible during the survey as follows:

Bed		Thickness m
10	Marlstone Rock Bed: Limestone; orange brown, ferruginous, sandy, shelly; green when fresh, rubbly and flaggy	3.00
9	Marlstone Rock Bed: Sandstone; poorly cemented, soft, fine-grained micaceous, ferruginous; small rounded pebbles of sandstone and siltstone at base	0.50
8	Siltstone; with sandy ferruginous and calcite veins; shells; belemnites near base	0.64
7	Siltstone; grey, poorly cemented, ochreous, micaceous; iron-stained fractures and surfaces	2.20
6	Sandstone; orange, poorly cemented, calcareous; abundant shells; iron-stained fractures (Bed 4 of Howarth, 1958)	0.54
5	Mudstone; pale grey, silty, with sandy wisps; ferruginous sandstone doggers up to 1 m across	2.20
4	Limestone; grey-green, weathering reddish orange, sandy, ferruginous; calcite veins; shelly cementstone doggers	0–0.88

		Thickness m
3	Mudstone; pale grey to brown, silty, finer towards base; shell debris; ironstone nodules (partially obscured by scree)	4.8
2	Sandstone; poorly cemented, with patches of silt; abundant doggers up to 1.5 × 5 m of medium-grained calcareous sandstone, weathering to orange to green-brown sand; bioturbated pockets of shells; shell debris; calcite-lined burrows filled with sandstone; pectinids and other bivalves; crinoid debris including *Pentacrinites*; dark grey phosphatic nodules weathering red-brown (Bed 2 of Howarth, 1958)	1.5
1	Mudstone; pale grey, weathering brown; a little silt towards top, more towards base; ferruginous bands and small doggers common in top 1.0 m	4.0

Fossils collected included: from Bed 4; *Pholadomya sp.*, *Pleuromya?*, *Protocardia truncata*, *Amaltheus* cf. *subnodosus*; from Bed 6, *Camptonectes sp.*, *Ceratomya sp.*, *Modiolus scalprum*, *P. truncata*, *Pseudolimea sp.*, *Pseudopecten sp.*, *Unicardium cardioides*, *Amaltheus sp.*; from Bed 10, *Camptonectes mundus*, *Modiolus scalprum*, *P. truncata*, *Pseudolimea sp.*, crinoid fragments. Mr R. H. Hoare collected in 1951, *Amaltheus* cf. *margaritatus*, *A. stokesi* and *Arieticeras* aff. *nitescens*. Thus both the two lower subzones of the *margaritatus* Zone are proved but not the highest subzone of *A. gibbosus*. Large doggers similar to those of Bed 2 were noted by Edmonds and others (1965, p.44) at Priors Marston Brick Works [493 575].

Silts, sandy silts and fine sands cap three hills near Upper Shuckburgh [4900 6100; 4944 6157; 4960 6200]; the most northerly shows a little brash of shell, detrital, ferruginous limestone. Up to 4 m of ferruginous silts crop out on the top of a small hill [5053 6030] SW of Lower Farm. Bush Hill [5100 6348], Flecknoe, is capped by about 11 m of silts and clayey silts; a temporary trench hereabouts [5126 6334 – 5131 6320] exposed about 4 m of micaceous silts with ferruginous nodules and sandy wisps. JB

Hellidon to Bates Farm

This area takes in the main escarpment. A small exposure [5173 5850] north of Hellidon church shows 0.3 m of ferruginous shell detrital limestone overlying about 5 m of micaceous silt and siltstone, the basal beds of the Middle Lias. In a nearby stream section [5186 5855] the transition from pale grey slightly silty clay (Lower Lias) to micaceous silts (Middle Lias) takes place within less than 1 m of strata.

At the northern end of the outcrop a small feature [5347 6117] about 20 m above the base of the Middle Lias marks a thin, sandy, ferruginous, shelly, detrital limestone. About 5 m above, and 5 m below the Marlstone Rock Bed, a stronger more persistent feature is formed by another similar limestone up to 0.5 m thick; weathered fragments of this bed show thick limonitic coatings and closely resemble material from the Marlstone Rock Bed. JB

Barby Hill

Few exposures occur within the district on Barby Hill. Core recovery was poor in the top 15 m of Barby Borehole, and the base of the Middle Lias is taken at 9.6 m depth by reference to the gamma ray log. Fragments of core include pale grey, micaceous siltstone, some with much iron staining, fine-grained sandstone and, below about 6.5 m, nodular orange-brown to dark purplish brown, ferruginous sandstone with scattered ooliths. Bioturbation is evident in some fragments, and a little plant, wood and shell debris

is present. Lenticular masses of ferruginous sandstone noted just east of the district resemble the nodular sandstone in the borehole. Pale grey silts and silty clays, weathering red and ochreous, with lenticles of ferruginous siltstone and sandstone, occur on the northern slopes of Barby Hill. KA,MGS

A ferruginous sandstone caps Barby Hill [538 695], and dips gently to the ESE. This bed has been regarded by previous workers as the Marlstone Rock Bed, but it does not resemble that bed in its lithology, showing a variety of lithological types (although all are ferruginous) and a lack of its usual brachiopods. KA

A small quarry [5285 5933] in the Marlstone Rock Bed, now filled, at Upper Catesby, showed (Whitehead and others, 1952):

	Thickness m
Upper Lias (see below)	0.26
Oolitic stone, hard, dense, blue-green, ferruginous; limonite staining in joints; numerous belemnites	0.23
Ferruginous oolitic stone; shelly lenses; strings of iron-pan parallel to bedding planes and joints; '*Rhynchonella*' common	0.61

The iron content of the rock was 22.4 per cent. Bivalves found in the Marlstone Rock Bed at Upper Catesby [5264 5935] included *Grammatodon sp.*, *Protocardia truncata* and *Pseudolimea sp.*. JB

UPPER LIAS

Upper Lias strata are restricted to the south-eastern extremity of the district and comprise 2 – 3 m of interbedded limestones and paper shales overlain by up to 36 m of pale grey mudstone. Howarth (1978) has subdivided the basal Upper Lias of Northamptonshire as follows:

Upper Cephalopod Bed: brown ferruginous limestone
Clay; grey to brown, ferruginous in parts
Lower Cephalopod Bed: pale yellow-brown limestone, variably sandy, oolitic or shaly
Clay; blue-grey
Abnormal Fish Bed: pale blue-grey limestone, oolitic in parts
Transition Bed: pale brown oolitic ferruginous limestone

The Transition Bed, 0.15 m thick, has almost certainly been mapped as part of the Marlstone Rock Bed. Both of the Cephalopod Beds and the Abnormal Fish Bed contain characteristic ammonite faunas of the *bifrons* and *falciferum* zones. No Upper Lias strata younger than *bifrons* Zone age are known in the district.

The Abnormal Fish Bed corresponds to the later part of the *exaratum* Subzone, the Lower Cephalopod Bed lies within the *falciferum* Subzone and the Upper Cephalopod Bed within the *commune* Subzone (Table 3). Fossils found during the recent survey include: *Dactylioceras commune*, *Harpoceras falciferum*, *H. serpentinum*, *Hildoceras sublevisoni*, *Nodiocoeloceras*, indeterminate belemnites and rhynchonelloid fragments.

A major transgression followed deposition of the Middle Lias. Early sedimentation in the Upper Lias gave rise to interbedded limestone and bituminous shale, which Hallam and Bradshaw (1979) have interpreted as marking the maximum spread of anaerobic conditions in a topographic low on the sea floor; most of the Upper Lias strata are mudstones similar to those of the Lower Lias. KA

Details

An old quarry [5285 5933] at Upper Catesby showed, above the Marlstone Rock Bed (Whitehead and others, 1952):

	Thickness m
ABNORMAL FISH BED	
Limestone; flaggy to nodular, marly with scattered ooliths, brown to cream coloured, iron-stained in parts; *Harpoceras glyptum*, *Hildoceras* aff. *gyrale*, belemnites	0.15
(TRANSITION BED)	
Sandrock, ferruginous up to	0.03
Oolite; cream coloured or greenish, dark green where unweathered, non-ferruginous	0.08

JB

NORTHAMPTON SAND FORMATION

The lowest division of the Middle Jurassic, the Northampton Sand Formation, rests unconformably on the Upper Lias. It occurs within the present district as one small outlier at Studborough Clump [5341 5965], where friable, medium-grained ferruginous sands cap the hill top to a depth of about 2.5 m. Tilting, uplift and erosion of the Upper Lias (Hallam *in* Sylvester-Bradley and Ford, 1968, p.199) preceded deposition of the Northampton Sand. This ferruginous sandstone facies passes into oolitic ironstones to the east and north-east; a depositional environment comparable to that of the Marlstone Rock Bed is envisaged.

CHAPTER 7

Quaternary

CLASSIFICATION AND NOMENCLATURE

The town of Warwick lies just beyond the western extremity of an extensive spread of glacial drift that stretches north-eastwards to Grantham and south-eastwards to Cambridge, and extends in a tongue from Rugby south-westwards to Royal Leamington Spa (Figure 18). Sands and gravels that preceded the glaciation are preserved locally, and flights of river terraces were laid down along the main streams after the ice melted. Outside this area, glacial drift is confined to small outliers, the largest of which, near Burton Green in the north-west, is an extension of a larger tract of glacial drift around Birmingham. Other less extensive and more recent drift deposits include head, peat and alluvium.

Within the main outcrop of the glacial deposits there is a clear stratigraphy which was first recognised by Shotton (1953) and has been largely confirmed by the present survey. Minor changes have been made to Shotton's original nomenclature, largely to take account of later work (Shotton, 1976a; Rice, 1968, 1981; Douglas, 1974, 1980; Sumbler, 1983a). The nomenclature used in this account, and that originally employed by Shotton, are set out in Table 4. The oldest proved drift in this area is the Baginton Sand and Gravel, laid down by streams on the pre-glacial surface. The overlying Thrussington Till marks the onset of glacial conditions. It is derived largely from the Trias, having been deposited by ice moving from the north and north-east. It is overlain by Wolston Clay, consisting of laminated clays and silts with intermixed lenses of till. Sand and gravel intercalated in the Wolston Clay has been called the Wolston

Sand and Gravel. Near Rugby the Wolston Clay is replaced by Oadby Till, a chalk- and flint-bearing till of easterly derivation. Dunsmore Gravel, which caps the glacial sequence, is the outwash deposit of the Oadby Till ice sheet.

This sequence has assumed national importance since being proposed as the type for the Wolstonian Stage (Shotton and West, 1969), which covers the deposits laid down during a glacial episode that is believed to have intervened between the preceding Hoxnian and the subsequent Ipswichian interglacials. There is, however, no firm evidence that the type-deposits occupy this stratigraphic position, for as yet no Hoxnian or Ipswichian deposits have been found in sequence with the glacial deposits. The Baginton Sand and Gravel has yielded a cold-climate fauna that is believed to be post-Hoxnian (Shotton, Banham and Bishop, 1977; Shotton, 1984), but some of the evidence is not wholly conclusive (Sumbler, 1983a, b). In 1985 a warm temperate fauna and flora was obtained from a channel beneath these cold-climate gravels, but as yet they have not been dated. Ipswichian faunas have been recorded from the 3rd Terrace of the Stour, a tributary of the Avon at Ailstone some 10 km south-west of Warwick, and from an isolated outcrop near Coventry (Shotton, 1929); the patches of 3rd Terrace in the Warwick district are, however, so small, so few, and so isolated from one another, that their correlation with the above deposits can be only speculative, though it is generally agreed that the faunas post-date the Wolstonian deposits. Because the Oadby Till is chalk-bearing, and because the on-

Table 4 General stratigraphy of the glacial drift: comparison of present and earlier nomenclature

Shotton (1953)	(Shotton (1976a)		This memoir	
Dunsmore Gravel	Dunsmore Gravel		Dunsmore Gravel	
Upper Wolston Clay	Upper Oadby Till Lower Oadby Till		Upper Wolston Clay	Oadby Till
Wolston Sand	Wolston Sand		Wolston Sand and Gravel	
Lower Wolston Clay	Bosworth Clays and Silts		Lower Wolston Clay	
	Thrussington Till		Thrussington Till	
Baginton Sand	Baginton Sand		Baginton Sand — — — and — — — Gravel	
Baginton-Lillington Gravel	Baginton-Lillington Gravel			

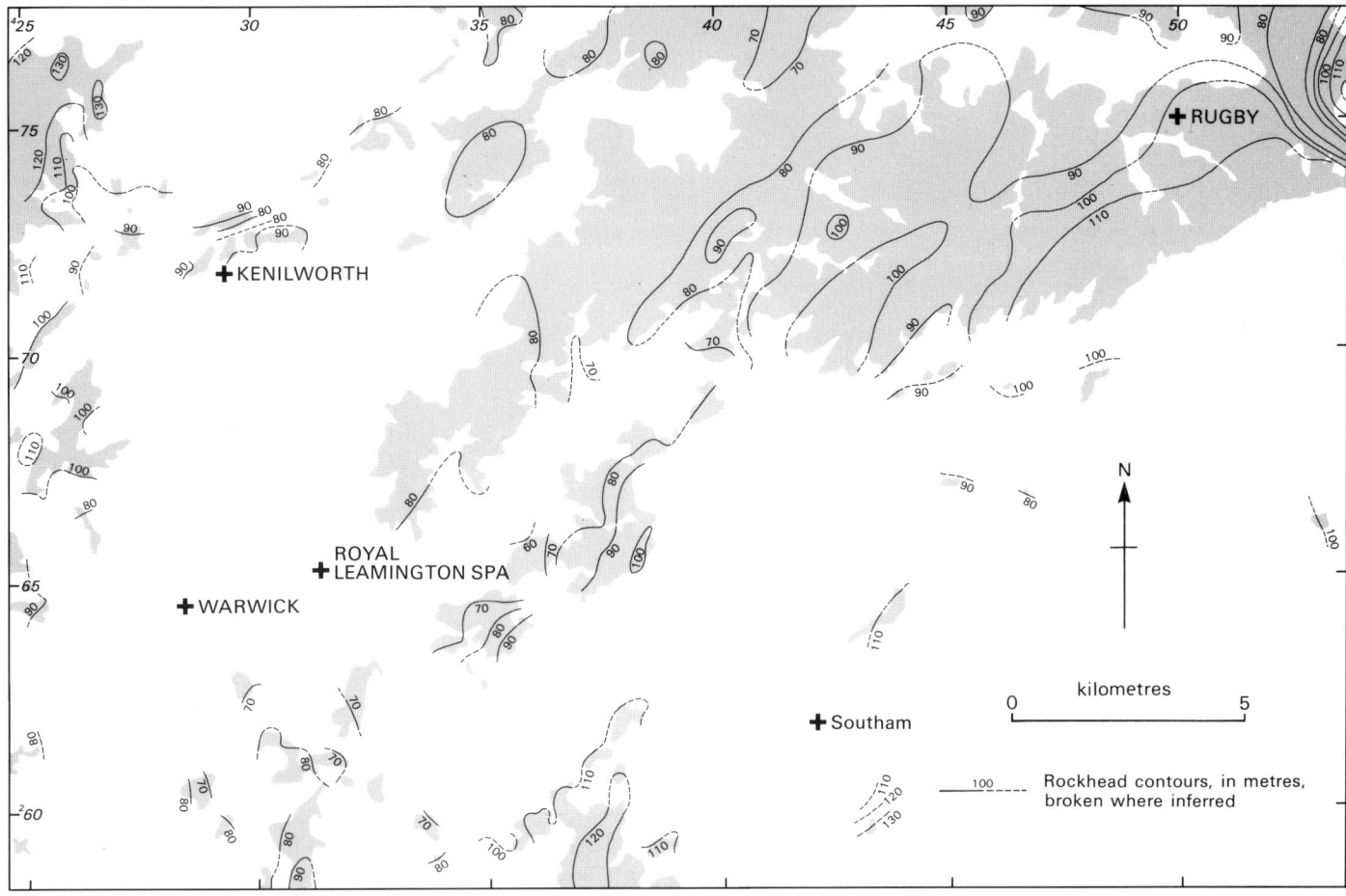

Figure 18   Sub-drift bed-rock topography of the Warwick district

ly chalk-bearing tills yet dated with certainty in eastern England are Anglian in age, the present authors prefer to regard the glacial drift of the Warwick district as Anglian (Sumbler, 1983a), though recognising that this correlation is as tenuous as the conventional one. The term Wolstonian has, however, been retained in this account, though its usage is subject to the above proviso.

The outliers of glacial deposits away from the main outcrop become progressively more difficult to correlate with the standard sequence as their distance from that outcrop increases. They are treated as the products of the same Wolstonian glaciation, and are described below under the stratigraphic headings that seem to be the most appropriate.

BAGINTON SAND AND GRAVEL

A reconstruction of the rock-head surface beneath the district is shown in Figure 18 and another has been given by Crofts (1982, figure 2). There is some evidence for the presence of a rock-head depression running roughly beneath the present valley of the Avon upstream from Royal Leamington Spa. Shotton (1953) has interpreted this depression as the pre-glacial valley of the 'proto-Soar' which followed a north-eastward course from near Evesham to Leicester and beyond.

Within the Warwick district it is hard to evaluate this contention, for not only are data meagre but the rock-head contours are the integration of several distinct episodes of subaerial, glacial, and possibly subglacial drainage, and so are most unlikely to represent the surface topography at any given time.

Within this rock-head depression, Shotton (1953, pp.213–214) has described clays at the base of the local drift sequence near Bubbenhall [358 718] and at Whitley [352 769]. He named the deposit the Bubbenhall Clay, and regarded it as Anglian in age and possibly lacustrine in origin. During the survey several boreholes were drilled to investigate the Bubbenhall Clay, but it was not found in any of them, nor was it located at outcrop (Sumbler, 1983a).

The oldest Quaternary deposit known with certainty is the Baginton Sand and Gravel which is preserved extensively beneath boulder clay from Wolston to Royal Leamington Spa and crops out widely on both sides of the Avon, especially at Baginton and up-river from Bubbenhall. From Coventry through Kenilworth to the western margin of the district, there are several elongated outliers of sand and gravel which have also been assigned to this unit. The term has been introduced to include both the Baginton-Lillington Gravel and the Baginton Sand of Shotton (1953) for, although sands normally overlie and overstep gravels within the deposit, in many places there is no mappable boundary between the two lithologies. The maximum thickness between Bubbenhall and Ryton-on-Dunsmore is about 10 m, the gravel and the sand each being some 5 m thick.

The gravel is of two types (Shotton, 1953, pp.214–216). The 'Baginton facies' is dominant north of Weston under Wetherley [366 695]. It consists almost entirely (up to 98 per cent) of well rounded pebbles of quartzite and quartz (hereafter, Bunter pebbles or quartzites), presumably derived from the Triassic 'Bunter' Pebble Beds , with a little locally derived Triassic material, particularly near the base. Coal is locally abundant on foresets. The outliers between Baginton and Kenilworth also contain a little flint. The 'Lillington facies' occurs south of Weston under Wetherley. 'Bunter' pebbles are again predominant, but are accompanied by Jurassic limestones, ironstones and robust fossils. In both facies the composition of the clasts is hard to reconcile with the postulated direction of drainage of the proto-Soar, which is claimed to have laid down the deposits (Shotton, 1953). The Jurassic material probably came from the Cotswolds, and the 'Bunter' pebbles and coal ultimately from the West Midlands, although the clasts may have been reworked from Tertiary or older Quaternary deposits of which no evidence remains.

A fauna including *Coelodonta antiquitatis*, *Mammuthus primigenius* and *Equus caballus* has been obtained from the base of the gravel at Lillington, and *E. caballus* has been found also at Baginton: a cold steppe environment is implied (Shotton, 1953, pp.219–220). Silts within the deposit at Brandon have yielded fish, insects, and a flora including *Betula* and *Salix*, all similarly pointing to a cold, even subarctic climate (Shotton, 1953; Kelly, 1968; Osborne and Shotton, 1968), though they are not diagnostic of any individual stage. A single footbone of *Coelodonta antiquitatis* from Manor Pit, Lillington, is evidence that the gravel is no older than the Hoxnian, and is the best evidence of the age of the overlying 'Wolstonian' (Shotton, 1984).

The overlying sand is generally fine- to medium-grained, and is clean and well sorted, with many lenses of 'Bunter' pebbles and coal fragments, particularly in concentrations along the foresets. Current directions, as indicated by the foresets, are variable. In the pits at Bubbenhall, Ryton-on-Dunsmore and Wolston they mostly suggest a flow from between west and south.

Details

Outliers west of Sowe

From the western margin of the district to near Pinley there is a chain of small disconnected outliers, mostly aligned NNE–SSW; in many of these the Baginton Sand and Gravel appears to fill channels cut into Carboniferous and Triassic strata. The three westernmost outliers extend in a narrow strip for 1.6 km. A well [2569 7078] at Fernhill Farm into one of these proved 12 m of sand and gravel above Bromsgrove Sandstone.

Other outliers occur on both sides of Finham Brook at Kenilworth. Exposures are poor south of the brook, but on its north side at The Common [299 730] the channel floor slopes down steeply to the north, and 10 m of gravel are exposed in the railway cutting [2983 7315]. In Cherry Orchard brickworks [295 722], 11 m of coarse well bedded brown sand with a little sandy gravel lie beneath Oadby-type till; some beds are composed almost entirely of com-

minuted coal, while the larger clasts are mostly Bunter quartzites though there are also angular Enville Group sandstones and mudstones. Cross-bedding is from the south, almost at right angles to the trend of channel. Farther east, excavations for a gas-main [3132 7396] proved 2 m of sandy and silty gravel containing Bunter quartzites and a few flints; large boulders of sandstone and a few of limestone lay near rock-head. North of King's Hill, a borehole [3270 7501] in another of the outliers proved 9 m of sand and gravel; these deposits were classified as the Fourth Terrace of the Sowe by Shotton (1953, p.288), but they seem too high for this to be correct, and contain only scattered flints although these are generally abundant in the terrace gravels. In the north of this tract a temporary section [353 770] south-west of Pinley was described by Shotton (1932) as 4 m of fine buff loamy sands with a few 'black bands' (presumably coal) and interbedded seams of Bunter-rich gravel. The bedding was greatly disturbed and contorted; Shotton ascribed this to overriding by Wolstonian ice. In 1980, a nearby excavation [3542 7702] began in Wolston Clay, passed through 1.5 m of coarse Bunter-rich gravel containing some flints, and entered 2.5 m of pink to brown, cross-bedded sand with much coal on the foresets; the bedding was disturbed, and the foresets now dip at 40°.

RAO, MGS

Baginton, Bubbenhall and Wolston

The main outcrops lie in this area and the deposit has been worked extensively on both sides of the Avon. Around Baginton it forms a plateau with a surface level at about 85 m. The gravel is here about 5 m thick, though it locally ranges up to 8 m, but the overlying sand is seldom more than 2 m thick. Shotton (1929, pp.210–212) records a section of the workings in a pit [348 750] that has now been backfilled; beneath 1.2 m of till (Thrussington Till), there were 3.7 m of fine red sand with a few layers of pebbles (Baginton Sand), which rested on about 3.1 m of coarse red and yellow gravel (Baginton Gravel). Farther upstream, but still to the north of Avon, he described another gravel pit [383 762] west of Brandon Wood (Shotton, 1968, pp.393–395), in which the Baginton Gravel was cut by an ENE-trending channel about 30 m wide. The channel was filled with up to 2 m of purple-grey silty clay and brown laminated silt, the surface of which was intensely disturbed by cryoturbation. Wedges of similar material also occur in another pit nearby [394 763] (Shotton, 1968). A working pit [3871 7643] in this same spread has exposed 3 m of pink cross-bedded sand with coaly layers overlying 0.5 m of gravel with Bunter pebbles; the sequence is capped by Thrussington Till. East of Brandon, a disused pit [412 763] worked up to 3.7 m of the Baginton Sand and Gravel, which rested on Mercia Mudstone and was overlapped by river gravels ascribed to the Fourth Terrace (Shotton, 1968, p.390, fig. 3). Near the eastern limit of the deposit at Brandon Grange, 2 m of brown and pink sandy gravel have been exposed in a temporary excavation [4203 7690]; the gravel contains Bunter pebbles together with a wedge of purple clay and brown clayey sand. Similar clay has been dug from a nearby ditch [4208 7703].

Some 5 km east of Brandon, and just north of the district, sand and gravel at the base of the drift is exposed at the northern end of a disused clay-pit [465 777]. Shotton (1953, pp.217–219) considers that this deposit is part of the Baginton Sand and Gravel, and that it is the source of mammalian remains (see p.60) first described by Buckland (1823), but at least some may have come from nearby terrace gravels. Up to 2.7 m of red and orange sand are recorded as overlying gravel which rests on the Lower Lias; the gravel contains Jurassic limestones and ironstones together with Bunter pebbles. Only about 0.6 m of gravel is still visible and, while Jurassic material is common, no Bunter pebbles or flints were noted.

To the south of the Avon there are extensive workings between Bubbenhall and Ryton-on-Dunsmore, and smaller ones near Wolston. Ryton Wood Pit [3764 7261] is dug through Thrussington

Till into 4 m of pale red-brown medium-grained sand with comminuted coal on foresets; this sand rests on about 1 m of close-packed Bunter-rich gravel. Up to 3 m of the gravel, with intercalations of coarse sand, are exposed in another pit [3819 7485] northwest of Ryton and, at Manor Farm Pit [392 737], 3 m of sand with lenses of pebbles overlie 4 m of Bunter-rich gravel. Upstream, in Wolston Pit [410 747], which is now largely filled, 3 m of sand identical to that at the other pits lie beneath Thrussington Till, and Shotton (1953, pp.216–217) records 4.3 m of sand overlying 1.5 to 4.9 m of gravel, which rested on Mercia Mudstone.   MGS

Outliers south of Warwick

Several small outliers of sand and gravel occur south-west of a line from Hampton on the Hill to Whitnash. Their correlation with the Baginton Sand and Gravel is, at best, speculation, but is convenient in practice. There are four of these outliers west of the Avon between Budbrooke and Hampton Gorse. The deposits contain Bunter pebbles, some Jurassic pebbles locally, but few flints. Shotton (1953, pp.240–241) assigned them to the Lillington facies. East of the Avon, Watchbury Hill [287 605] and a hilltop south-east of The Asps [297 628] are both capped by coarse Bunter-rich gravel with a few scattered flints in a matrix of sand; at least 7 m of gravel are preserved at The Asps and perhaps 4 m at Watchbury Hill. Farther east at Tachbrook Mallory [316 621], a lens of yellow medium-grained sand, up to 2 m thick, is overlapped eastwards by Thrussington Till.   KA

THRUSSINGTON TILL

Rice (1968), working in central Leicestershire, has described a till, variously red and Trias-rich or grey and Lias-rich, that is almost completely lacking in Cretaceous erratics; he named it the Thrussington Till. A south-western lobe of this composite till-sheet enters the Warwick district and extends as far as Royal Leamington Spa along the low ground of the Mercia Mudstone outcrop, rarely occurring above 90 m above OD. Isolated outliers continue southward on the east of the Avon to the southern extremity of the district.

The till is generally 3 to 5 m thick, and it overlies and overlaps the Baginton Sand and Gravel. Its basal contact is generally flat or gently undulating (Plate 7) and sharp, suggesting that the underlying sand was frozen when the till was deposited. Locally near Weston under Wetherley [365 693; 368 705; 370 713] and Hunningham [370 670], however, the till cuts down through the Baginton Sand and Gravel for 5 to 10 m into the Mercia Mudstone, and at such localities there was clearly erosion before the ice melted, while the sand and gravel at Whitley is highly contorted (Shotton, 1932), possibly because of the overriding ice-sheet.

The contact with the overlying Lower Wolston Clay is gradational, and there is an upward transition, through about 1 m, from red stony till to grey-brown, largely stoneless clay. South of the Leam, though lenses of stoneless clay occur within the till, the Lower Wolston Clay is generally absent as a discrete formation, and the till is overlain by the Wolston Sand and Gravel.

The Trias-rich variant of the Thrussington Till is dominant. It is a tough red-brown clay containing pebbles and blocks mostly of green-grey Triassic sandstone and siltstone, with some of red mudstone and Carboniferous sandstone, many Bunter pebbles and small fragments of coal. Rarer erratics include Leicestershire diorites, Coal Measure ironstones and Lower Carboniferous limestones. In detached outliers between Eathorpe and Radford Semele, Jurassic material is common and there are rare flints. Both lodgement-till and flow-till are probably present within the Thrussington Till, and there are local lenses of almost stoneless, smooth clay, very like the Wolston Clay.

The bulk of the till has been derived from the Mercia Mudstone that underlay much of the ice-sheet, though the Bromsgrove Sandstone and Enville Group may have been contributory sources. The erratics establish that the general ice-flow was south-westward along the broad depression that follows the outcrop of the Mercia Mudstone. The Jurassic erratics south of the Leam may indicate an eastern tributary to a composite ice-sheet.

Details

Royal Leamington Spa to Wolston

Up to 0.7 m of red Thrussington Till overlie Baginton Sand and Gravel in an old pit [3290 6628] on Royal Leamington Spa golf course. Platy pebbles are aligned subparallel to the base in places. The basal few centimetres are mostly of stone-free clay, and clay lenses occur higher up. The base is irregular and shows some interlayered sand and clay. Temporary excavations [328 665] north of Campion Hill showed the sharp basal contact of the till; up to 0.3 m of dark, red-brown, smooth, plastic clay with rare pebbles occurred at the base of the till. Excavations at Weston under Wetherley [3662 6907] showed red, sandy, pebbly clay and pebbly sand, with pockets of pebble-free sand. The pebbles included Bunter quartzite, Triassic sandstone and siltstone, coal and ironstone, but no flint. However, flints are present in the till in an old pit farther west [3515 6980]. Cubbington (BGS) Borehole [3493 6917] proved 2.9 m of till and Weston under Wetherley Borehole [3618 6911] proved 3 m of till resting on Mercia Mudstone.   KA

Burnthurst Farm Borehole [3880 7158] (Sumbler, 1981) passed through Wolston Clay into 1.7 m of till consisting of red-brown clay with many pebbles of Triassic sandstone and siltstone and Bunter quartzite, and scattered coal specks. At the south end of Ryton Wood Pit [3746 7210] 3.5 m of typical Thrussington Till are very sandy in the basal 0.2 m. The junction with underlying Baginton Sand is sharp but undulating, and the till grades upwards into Wolston Clay. At the north end of the pit [3789 7271] 4 m of till, with large blocks of Triassic sandstone and siltstone and scattered pods of brown sand, underlie Wolston Clay; the junction with the underlying Baginton Sand is horizontal. In Manor Farm Pit [3933 7363], 3 m of till overlie Baginton Sand; trial bores to the south-west [389 735] indicate thicker till and thinner sand.

Wolston Pit [4105 7465] shows 3 to 4 m of locally sandy till beneath Wolston Clay. In the eastern part of the section the till grades into underlying Baginton Sand; in the western part the contact is sharp.   MGS

Radford Semele to Hunningham

Radford Semele Nos. 2, 3 and 4 boreholes each proved thin Thrussington Till at the base of the drift. Pebbles in red coaly till in an old pit [3548 6465] north of Leasowe Farm include some flint, in addition to the characteristic Trias-derived erratics. Offchurch No. 1 Borehole [3680 6545] proved a full thickness of 3 m of Thrussington Till. A cutting on the Fosse Way [3676 6478] exposes three distinct layers of till between the Baginton Sand and Gravel and the Wolston Sand and Gravel. At the base is 0.25 to 0.75 m of red-brown, very sandy, pebbly clay derived from Triassic rocks; this is overlain by up to 0.75 m of purplish brown to chocolate-brown,

Plate 7 Thrussington Till overlying cross-bedded Baginton Sand and Gravel; Wolston Pit.   A13085

slightly silty, pebbly clay with sandy pockets. A lens of orange-brown silt separates these two layers at one place. The uppermost till consists of up to 1.3 m of reddish brown, sandy, pebbly clay; it grades into the overlying Wolston Sand and Gravel. Lenses of brown, clayey, cross-bedded, pebbly sand occur between the middle and upper layers of till. All three layers contain Jurassic limestone and ironstone erratics in addition to the usual Triassic and Carboniferous rocks, though they are most abundant in the middle layer.

At the reservoir [3740 6528] south of Burnt Heath Farm, 2 m of Thrussington Till are exposed. It consists of dark red to chocolate-brown pebbly clay, mottled red, brown, grey and green, with carbonaceous streaks and red and green sandy pockets. Small erratics include Triassic siltstone and mudstone, Jurassic mudstone and ironstone, coal, and sporadic Bunter quartzite. Larger erratics, up to 0.3 m across, include Jurassic ironstone, shale and limestone (including Lower Lias cementstone and fossils), Langport Member limestone, Carboniferous sandstone and limestone, and mica schist.

Hunningham Borehole [3761 6724] proved 8.8 m of till, with interbedded sand, below the Wolston Clay. The uppermost 5.4 m were red and Trias-rich, with few Jurassic erratics; the lower part included brown till containing more Jurassic material.   KA

WOLSTON CLAY

The Wolston Clay is present in the main area of glacial drift, and may be represented by similar clay in several of the outliers. In many places it comprises lower and upper divisions separated by the Wolston Sand and Gravel.

The dominant lithology is an almost stoneless clay or silty clay, grey-brown to chocolate brown when fresh, but weathering to reddish brown; it is usually calcareous, particularly in the upper part, and commonly contains small nodules of race in the weathered zone. The Wolston Clay, though at first sight massive and structureless, is generally finely laminated; laminae and lenticles of pale brown silt and sand occur, especially just below the Wolston Sand and Gravel. These laminated beds indicate deposition in standing water. They may comprise annually accreted varves (Shotton, 1976a, p.248), but Douglas (1980, p.282) maintains that similar laminae in broadly equivalent deposits in Leicestershire are small-scale turbidites, citing fining-upward layers and cross-lamination as evidence.

The Wolston Clay is rarely completely free of stones, and usually contains scattered pebbles interpreted by Shotton (1953, p.223) as drop-stones from icebergs melting in a glacially impounded lake. The most common pebbles in its lower part are of Bunter quartzite and Triassic sandstone and siltstone (as in the Thrussington Till). Chalk, flints and Jurassic limestone are commoner in its upper part marking the arrival in the area of the Oadby Till ice-sheet. Larger erratics, many polished and striated, including Leicestershire granodiorites, Jurassic limestone, and Carboniferous limestone and sandstone are common over the outcrop. Bodies of stony till occur locally within the Wolston Clay, particularly in its upper part, and this till facies predominates around Cubbington, Radford, Hunningham and Harbury, and in the western part of the district. Some of the tills are red and mainly of Triassic material; others are brown or grey and derived from Chalk and Jurassic rocks; the two types resemble the Thrussington and Oadby tills respectively.

Between Royal Leamington Spa and Brandon, the Wolston Clay rests on the Thrussington Till, and is up to 18 m thick around Princethorpe and Bubbenhall. It overlaps the Thrussington Till, and thins progressively to the east and southeast against the rising subdrift surface. The Lower Wolston Clay is overlapped by the Wolston Sand and Gravel at Hunningham and Thurlaston. At Radford Semele, and between Thurlaston and Rugby, the Upper Wolston Clay is in turn overlapped by the Dunsmore Gravel, deposition of the gravel having been preceded and accompanied by some erosion. In the deep drift-filled valley east of Rugby, the Wolston Clay is at least 30 m thick, and possibly 50 m (Figure 18); it passes laterally, and vertically upwards, into Hillmorton Sand. Farther east, and to the north of the district, the Wolston Clay is replaced laterally and is overlain by Oadby Till (Sumbler, 1983a).

Because lacustrine deposits like the Wolston Clay are widespread here and elsewhere in the Midlands, Shotton (1953) suggested that they were laid down in a vast glacial Lake Harrison, thought to have covered the central Midlands from Market Bosworth and Leicester in the north to Moreton-in-Marsh in the south, and to have been bounded to the north and south-west by ice-sheets, to the west by the high ground around Birmingham and Redditch, and to the south-east by the Jurassic escarpment. He suggested a water-level of about 125 m above OD: Bishop (1958) suggested 133 m. The evidence for this single extensive lake is far from convincing as there is no reason to suppose that the lacustrine deposits are everywhere precisely contemporaneous. Furthermore Dury's (1951) supposed lake-shoreline feature on the Jurassic scarp is formed in the present district by the

outcrop of the '70-Marker' Member, a hard bed in the Lower Lias (Ambrose and Brewster, 1982). The close association between the Wolston Clay and the overlying and underlying tills, coupled with the widespread occurrence of till in the Wolston Clay, suggest the close proximity of ice. Possibly the clays accumulated in transient glacial lakes and ponds formed in front of, upon, and even within ice-sheets advancing intermittently from the north and east (Sumbler, 1983a).

Details

North of the Avon

Laminated clays enter the district from the north at several places between Pinley and the outskirts of Coventry. In a section [3542 7702] near Whitley Hospital, 1.1 m of red-brown, smooth, sticky clay, with rare pebbles of Bunter quartzite and Triassic mudstone and siltstone, overlies 0.4 m of soft red silt and Baginton Sand and Gravel. A little to the east, Brandon Wood Pit [3871 7644] shows 1 m of brown plastic clay which passes down into Thrussington Till. Farther east the Thrussington till is overlapped by Wolston Clay and, in an old gravel pit [4032 7670], 3 m of ochreous-weathering, dark red to blue-grey mottled clay, containing scattered pebbles of Bunter quartzite, Triassic siltstone and rare chalk, rest on Baginton Sand and Gravel.

Bones of bison and reindeer, and possibly bivalves, were recovered from clays overlying the Langport Member in a quarry [?4582 7728] near King's Newnham (Wilson, 1876; Oldham, 1879). The clay may have been Wolston Clay but, if so, this is the only record of a fauna from the unit and it is possible that it was part of a terrace deposit. Red, brown and grey, silty and sandy clay at the entrances to canal tunnels at Newbold on Avon contain rare pebbles of chalk, flint and Bunter quartzite [4865 7712], and Bunter quartzite, chalk and Triassic siltstone [4870 7737]. Geological Survey field maps surveyed in 1915 record brown clay with Bunter quartzite pebbles and a large block of Carboniferous limestone overlying Lower Lias in a quarry at Newbold on Avon. MGS

Royal Leamington Spa to Rugby

South of Weston under Wetherley, the Lower Wolston Clay is a chocolate-brown, smooth, laminated stoneless clay, generally 2 to 4 m thick. It thins southwards and south-westwards of Cubbington, and was only 0.8 m thick in Cubbington Borehole. The Upper Wolston Clay is about 2 to 3 m thick; in Cubbington Borehole it was mainly red-brown to grey, unbedded stony clay.

Farther east, Burnthurst Farm Borehole [3880 7158] (Sumbler, 1981) proved 12.6 m of Lower Wolston Clay, which consisted mostly of smooth, brownish grey, plastic clay with laminae and layers of pale brown silt; there was also some silty clay and a few layers of fine sand in the upper part. The lowest 2 m contained scattered pebbles of Triassic mudstone, siltstone and sandstone, and graded downwards into Thrussington Till; the top 2.5 m, below the Wolston Sand and Gravel, contained many pebbles of Bunter quartzite and Triassic siltstone, and also small nodules of calcareous race, which possibly imply subaerial weathering of the Lower Wolston Clay prior to deposition of the sand and gravel. The Upper Wolston Clay was 3.2 m thick. It was a blue-grey laminated clay with silty layers; pebbles of rare Bunter quartzite and Triassic siltstone, and much chalk occurred in the basal 0.3 m.

In a road cutting on the Fosse Way [3988 7021], 2.5 m of clayey pebbly till, whose erratics included a boulder of granodiorite from Leicestershire, rest on brown, bedded silt and clay. The Wolston Clay capping hills near Princethorpe [406 705; 408 701] is largely till-like, and includes much limestone from the Langport Member. Wolston Pit [4105 7465] exposes 1.35 m of dark purple-brown mottled clay with laminae of pale brown silt, especially in the upper part, and sporadic silt lenticles up to 5 mm thick; the basal 0.15 m is a smooth, red clay with silt layers, passing down into Thrussington Till.   MGS

A full thickness of up to 12.9 m of Wolston Clay (including up to 2.7 m of Wolston Sand and Gravel about 4 m above the base) was proved in boreholes at Blue Boar road-bridge [4534 7195]. It consisted of red-brown to grey-brown, sandy and silty clay with scattered pebbles and, in the Upper Wolston Clay, lenses of sand.

Several exposures and excavations in the Wolston Clay on its broad outcrop south-east of Wolston show its variable lithology. They include brown-weathering, blue-grey clay, in part laminated, containing rare blocks of Lower Lias cementstone and pebbles of Bunter quartzite, flint and chalk [4414 7269]; blue-grey and red-brown, poorly laminated, pebble-free clay [4434 7334]; smooth, dark blue-grey and brown clay with rare Triassic siltstone pebbles [4511 7429]; and soft, smooth, grey-brown, stoneless clay with some silty lenticles [4585 7382].

The clay is overstepped south-eastwards by the Dunsmore Gravel. Very locally it emerges from beneath cover on the south of the spread of Dunsmore Gravel, as in an old pit [4729 7125] which showed (Shotton, 1953, p.244) 0.9 m of red clay with some Bunter and flint pebbles, but almost stoneless at the base, separating the Dunsmore Gravel and Lias. A similar succession exposed [4724 7120] during construction of the M45 motorway included 0.6 m of blue-grey and purple-red mottled clay containing flints, Bunter quartzite and Triassic sandstone.

Between Cawston and Rugby the Wolston Clay contains a good deal of chalk. Boreholes [4873 7297] have proved Dunsmore Gravel on up to 3.4 m of grey and brown clay with chalk fragments and with layers of silty clay and sand in the upper part. Temporary sections showed 3 m of red-brown and grey mottled clay with sandy lenses and sparse pebbles including chalk, resting on Lower Lias [5002 7305], and 1 m of pale grey and red-brown mottled clay with race and scattered flint, chalk and Bunter pebbles, resting on Lias [5009 7460]. The Dunchurch Road Pit [502 744], now built over, showed 2.1 m of Wolston Clay beneath Dunsmore Gravel and resting on Lower Lias (Lowe, 1873). Wilson (1870a; 1870b) described the Wolston Clay here as stiff, brown clay containing commonly striated pebbles, mainly of chalk. Five boreholes at Rugby sports centre [5085 7448] proved 2.7 to 8.2 m of Wolston Clay beneath Dunsmore Gravel. The Wolston Clay comprised stiff, grey, silty clay with chalk pebbles, except that in one borehole this was overlain by 4.4 m of soft, brown and grey, silty clay with silt layers and with sand partings near the base.

Thompson (1899) recorded the Rugby cutting of the Great Central Railway, which showed up to 4.9 m of Wolston Clay underlying Dunsmore Gravel and resting on Lower Lias. At the southern end of the section [5154 7353] the Wolston Clay consisted of blue boulder clay (reworked Lias) containing chalk, flint and Bunter pebbles, and fossils and blocks of Lower Lias limestone overlain by chalky boulder clay containing lenses of gravel and sandy layers. In the northern part of the cutting the Wolston Clay consisted of blue-grey to brown clay with layers and laminae of sand, commonly contorted and containing a very few flint and chalk pebbles: this clay probably overlies the chalky clay. At Lanchester Polytechnic [515 749] boreholes passed from Dunsmore Gravel into 10.5 m of grey-brown, layered and laminated, silty clay and silt, locally sandy in its upper part. Farther east, the Wolston Clay grades into the Hillmorton Sand.

Foundation trenches [4850 7507] on the outskirts of Rugby showed 1.5 m of red and grey mottled clay with sandy patches and rare pebbles of Bunter quartzite, Triassic siltstone and Lower Lias cementstone. Boreholes in Rugby showed: 1.3 to 2.6 m of stiff, grey-brown, silty clay with scattered pebbles (Upper Wolston Clay), beneath Dunsmore Gravel and overlying Wolston Sand and Gravel [4990 7520]; 4.6 m of Lower Wolston Clay consisting of red-brown and blue-grey, silty clay with a few pebbles [4988 7541];

Dunsmore Gravel on 1.9 m of Upper Wolston Clay and 2.4 to 4.8 m of Lower Wolston Clay, both divisions comprising stiff, red-brown and grey mottled clay and silty clay with pebbles [502 753].

The railway cutting east of Rugby station [5176 7569 to 5239 7559], as described by Wilson (1870a, 1870b, 1874), exposed 15 m of Wolston Clay beneath Dunsmore Gravel. The clay was unstratified reworked Lias, and contained chalk, flint, quartzite, Lias limestone, red mudstone and sandstone. Augering during the recent survey, however, suggested the presence of the brown and grey, stoneless, silty clays. A borehole at Butler's Leap [c.5209 7589] (Wilson, 1870a, 1870b) passed from alluvium into sand, and then into clay with chalk pebbles; it may have reached the Lias at 16.8 m depth.   MGS

Radford Semele to Eathorpe

Laminated clays are less common in this area, and much of the deposit classified as Wolston Clay could well be termed till. The Lower Wolston Clay of Hunningham Borehole [3761 6724] comprised 1.1 m of smooth, chocolate-brown clay with ochreous and red-brown silt laminae and lenses. The Upper Wolston Clay of this area is dominantly till-like, but local, subordinate, stone-free, laminated clays or silts were proved in boreholes at Radford Semele, Offchurch, Hunningham and Fosse Farm. Red Triassic-rich till was proved in Offchurch No. 1 [3680 6545] and Fosse Farm [3768 6638] boreholes; in the latter it was 5 m thick and overlay 2 m of brown, Jurassic-rich till. The Upper Wolston Clay of Hunningham Borehole was mainly of Triassic material, but included 0.7 m of chalk-rich till with flints and Jurassic pebbles.   KA

WOLSTON SAND AND GRAVEL

The Wolston Clay is divided into lower and upper parts by a bed of sand and gravel over much of the district, although only the broadest outcrops have been shown on the published map. These deposits have been classified as Wolston Sand and Gravel, though they lie at different levels and contain different clasts. At Brinklow [440 790], just north of the district, there are two beds of sand separated by about 10 m of clay (Sumbler, 1983a): it is probably the upper of these that forms the Wolston Sand and Gravel at Newbold and Brownsover, north of Rugby, but over most of the district it is more likely to be the lower.

Over its main outcrop, between Royal Leamington Spa and Rugby, the Wolston Sand and Gravel forms a sheet 1 to 3 m thick, falling in level from around 105 m above OD at Rugby to about 90 m near Bubbenhall. Between Radford Semele and Eathorpe, the Wolston Sand and Gravel is over 12 m thick and its base as low as 70 m above OD. The typical lithology is fine- to medium-grained red sand and silt, commonly containing layers of plastic clay and silty clay. A basal pebbly layer is present in many places, and generally contains Bunter quartzite, green-grey Triassic sandstone and siltstone and rolled pellets of Wolston Clay. Jurassic and Cretaceous clasts are rare, and presumably most of the material was derived from the Trias-rich ice to the north which deposited the Thrussington Till.

In contrast, north of Rugby, Jurassic limestones and chalk are common in the upper bed, and Triassic clasts are rarer. This bed can be traced eastwards into a coarse gravel (the Shawell Gravel of Sumbler, 1983a) which is clearly outwash from Oadby Till ice.

Details

Royal Leamington Spa to Rugby

Lillington Well [3323 6806] penetrated 1.4 m of sand at 2.9 m (Richardson, 1928, p.117); in Cubbington Borehole this bed was 1.6 m thick. North-east of Cubbington [348 689] it is split into two by a bed of Wolston Clay.

A roadside exposure [354 692] near Weston under Wetherley shows 2 m of Wolston Sand and Gravel ranging from red-brown, medium- to fine-grained sand, clayey in parts with scattered small pebbles, to coarse, grey to yellow sand; the pebbles are mainly of Bunter quartzite, with Triassic sandstone and siltstone, Carboniferous sandstone and rare flints.

In Burnthurst Farm Borehole [3880 7158], the Wolston Sand and Gravel consisted of 1.0 m of very fine-grained, red sand, slightly clayey in parts, with some Bunter and green Triassic siltstone pebbles in the basal 0.3 m (Sumbler, 1981). Trial boreholes in Ryton Wood showed 0.8 m of orange-brown fine sand [3756 7225], and 0.6 m of clay and sand on 1.8 m of brown sand [3784 7250]. An overgrown pit [4015 7213] near Stretton-on-Dunsmore exposes 1 m of pale orange to pink and red silt and very fine-grained sand.

The lowest 1 m of the Wolston Sand and Gravel, as exposed in Wolston Pit [4105 7465], consists of medium- to coarse-grained, red-brown sand, passing into sand with grey clay and silt layers in the basal 0.15 m; a layer of pebbles up to 2 cms across, mainly Bunter quartzite and Triassic siltstone, occurs at the base. The full thickness of the Wolston Sand and Gravel here is about 5 m (Shotton, 1953, fig. 10).

North of the Avon, trial pits near Willenhall have proved red and brown sand below 1.4 m of Wolston Clay [3678 7639; 3673 7634], and 0.6 m of red sand within Wolston Clay [3669 7644].   MGS

Thurlaston to Newbold on Avon

Boreholes at Blue Boar road-bridge [4534 7195] proved 1.2 to 2.7 m of Wolston Sand and Gravel, consisting of red, silty, fine- to medium-grained sand. Boreholes in Rugby showed: at least 4.4 m of brown, silty, fine- to medium-grained sand with layers of brown clay [4990 7520]; a full thickness of Wolston Sand and Gravel comprising 5.2 m of sand [4984 7545]; 2.4 m to 3.0 m of red-brown sand with clayey layers, gravelly at the base [502 753].

At the eastern end of a canal tunnel [4872 7739] at Newbold on Avon, 1.0 m of red-brown clayey sand with clay lenticles and scattered pebbles of chalk rests on Lower Wolston Clay. A site-investigation borehole [4983 7722] to the east proved the Wolston Sand and Gravel to comprise 3.6 m of loose, fine- to medium-grained brown sand, clayey near the base, with a seam of brown sandy clay in the lower part.   MGS

Bishop's Tachbrook to Eathorpe

Several outcrops of sand in this chain of outliers have been referred to the Wolston Sand and Gravel, though only seldom does the sand lie within a sequence of laminated clays as it does at Wolston.

In the south-west, red and yellow sand up to 4 m thick has been worked [3089 6091] south-west of Bishop's Tachbrook. Around Tachbrook Mallory [320 621] the sand contains scattered Bunter quartzite pebbles.   JB

In Southam Road Borehole [3505 6430], Radford Semele, 9.7 m of coarse, red sand with scattered Bunter quartzite pebbles and abundant coal fragments, especially in the upper part, separate the Oadby and Thrussington tills (Richardson, 1928, p.85). Offchurch No. 1 and No. 3 boreholes proved 3.5 and 7.2 m of Wolston Sand and Gravel respectively. Fields between Offchurch and Eathorpe are littered with Bunter quartzite and flint pebbles [367 661; 371 662; 384 682; 397 692]. Fosse Farm Borehole proved 2.5 m of

Wolston Sand and Gravel (not bottomed), the upper part coarser with Bunter quartzite and Jurassic limestone pebbles and rare flints. Hunningham Borehole proved 2 m of coaly sand.   KA

HILLMORTON SAND

At Hillmorton, south-east of Rugby, a north-west-trending channel cuts into Lower Lias, closely following the present course of Clifton Brook. It may be a deepened pre-glacial valley. The deposits in it consist of fine- to medium-grained cross-bedded sand, with scattered lenses of gravel containing flint and Bunter quartzite pebbles and rarer chalk and Jurassic limestone. Lenses of smooth silty clay, some laminated, also occur, particularly in the lower parts. To the north-west, the sand passes laterally and vertically downwards into Wolston Clay, and the mapped boundary between the two is arbitrary. Near the margins of the channel, the Hillmorton Sand is overlain by Dunsmore Gravel. Boreholes have proved sand down to 75 m above OD and possibly to about 60 m above OD (Thompson, 1898), and the deposits may be up to 50 m thick. The Hillmorton Sand was probably deposited by waters draining north-westwards from the Northamptonshire uplands, ahead of the advancing Oadby Till ice-sheet.

Details

A sand-pit [531 744] at Hillmorton showed Dunsmore Gravel on 12 m of cross-bedded, pinkish brown, quartz sand with several gravel 'washouts' (Bishop, 1958, p.299). In the upper part of the deposit there were a few seams of pebbles, including limonitic ironstone, flint, Bunter quartzite, rolled *Gryphaea* and a rolled block of (?Lower Lias) clay: a BGS field map dating from 1950 also records coal and green micaceous shale (?Triassic). The underlying sands included beds of laminated silty clay up to 0.6 m thick, and some thin bands of silt.

Another pit [5339 7402] exposed 14.3 m of sand containing a few thin carbonaceous layers, probably coal fragments, and scattered flint and quartzite pebbles; seams of clay with rolled *Gryphaea* occurred in the upper part of the section (Wilson, 1870a, 1870b). A degraded face [5340 7405] in the pit shows 1.3 m of fine- to medium-grained orange-brown sand with sporadic pebbles and lenticles of grey clay.

Hillmorton Lane Borehole [5325 7482] proved 2.8 m of made ground and head on 0.6 m of very sandy grey clay with sand streaks on 4.6 m of clean, dark brown-grey silt and fine sand (not bottomed). Wilson (1870a, 1870b) recorded two trial boreholes as showing: sand, stiff and clayey below, to a depth of 16.2 m (not bottomed) [?534 747]; clayey sand to 8.8 m, just outside the district [5372 7444].   MGS

OADBY TILL

The term Oadby Till was introduced by Rice (1968) to denote a chalk- and Jurassic-rich till around Leicester. It extends into the Warwick district, where it consists of grey clay, weathering brown, with abundant pebbles and boulders of chalk, flint and Jurassic limestone and sandstone, and rarer ones of Bunter quartzite, Triassic sandstone, siltstone and mudstone, Carboniferous sandstone and coal. The clay matrix was largely derived from Jurassic mudstones, and transport was from between NNE and east. The main outcrop within the present district lies north-east of Rugby, and forms the southernmost tip of a tract of Oadby Till that extends northwards to Leicester; detached patches of similar till occur near Royal Leamington Spa and Kenilworth.

Around Royal Leamington Spa the till overlies the Wolston Sand and Gravel and relationships north of the district show that it is the lateral equivalent of the Wolston Clay. The Oadby Till ice-front probably stood in the Rugby area during deposition of much of the Lower Wolston Clay, and deposition of the Thrussington Till in the west, and Oadby Till in the east may have been broadly contemporaneous (Sumbler, 1983a). After deposition of the Wolston Sand and Gravel by meltwaters, further advance of ice laid down Oadby Till in the northern part of the district whilst the Upper Wolston Clay was deposited farther south. The outliers of Oadby Till lithologies to the west suggest that the ice may have eventually covered the whole district. It was locally erosive, and at Whitnash, Lillington and Cubbington the Oadby Till cuts down through Upper Wolston Clay, Wolston Sand and Gravel, Lower Wolston Clay and Thrussington Till.

Details

Rugby

Exposures of Oadby Till north of the town are in continuity with more extensive ones in the district to the north. On the northern side of the Avon, the till extends as far west as Brownsover. To the east the till fills a depression cut into the Lias, a continuation of that at Hillmorton (Figure 18). Its base is well below the level of the Avon alluvium.

To the south-east, the Oadby Till outcrops on the slope rising to Clifton upon Dunsmore. Vicarage Hill Borehole proved 12.6 m of Oadby Till beneath Dunsmore Gravel (Sumbler, 1981). The top 0.9 m consisted of grey to fawn clay with red sandy streaks, and contained chalk, flint, Triassic siltstone and mudstone and small specks of coal. The main part of the deposit was a very stiff clay containing abundant pebbles of chalk (from sand grade up to 5 cm across), subordinate flint and Jurassic limestone, and rare Bunter quartzite, Triassic mudstone, siltstone and sandstone, Carboniferous sandstone, and Jurassic ironstone and shale. The lower part of the till was vaguely layered, with specks and powdery streaks of chalk.

South-west of Clifton upon Dunsmore the till appears to pass laterally into the Wolston Clay, but the exact relationships are uncertain.   MGS

Outliers north of the Avon

The large spread of till at Burton Green [26 75] is chalky, with abundant orange and red flints; it has accordingly been referred to the Oadby Till. Near the western boundary of the district, however, this chalky till passes westwards into a red-brown sandy till of Thrussington-type, with abundant Bunter pebbles but little or no flint and chalk. The two tills are apparently contemporaneous, and were probably laid down by confluent lobes of ice that entered the district from the west and north. A complex sequence of gravels, sands, silts and clays fills irregularities in the subdrift topography where the two tills abut south and south-west of Burton Green. The complex includes beds of stoneless brown clay and silt, rather like the Wolston Clay. The clasts in these deposits are mainly Bunter pebbles, but a few flints are included.

The boulder clay in the large outlier north of Budbrooke consists mainly of the Trias-rich facies [2531 6756; 2542 6744]. However, in the extreme north between Bulloak Farm [257 689] and Bannerhill Farm [262 696] it is grey with orange and purple mottling, and contains some chalk and flint, as in the small outcrop north-east of Budbrooke [265 662]. On this basis it is thought likely that it equates with the Oadby Till. Due west of Budbrooke, a sequence of brown laminated clay and red Trias-rich, but flinty, boulder clay is presumably of the same general age; it rests on an impersistent till with Bunter pebbles, which may represent the Thrussington Till.

Farther east, there are two much smaller patches at Kenilworth where, at Cherry Orchard Brickworks [295 722], 6 m of brown to grey-brown boulder clay, with chalk, flint, Bunter quartzite and coal pebbles, and with large blocks of Enville Group sandstone near the base, rest on Baginton Sand and Gravel.

Outliers south of the Avon

There are small outliers of Oadby-type till between Royal Leamington Spa and Weston under Wetherley; they lie some 15 km south-west of the exposures north of Rugby. Near Lillington and Cubbington, the till rests on Lower and Upper Wolston Clay respectively, but eastwards and south-eastwards it cuts down to the Thrussington Till. It is rarely thicker than 1 m, the maximum recorded being 3 m.

Oadby till is also present in the outliers of drift north-east of Radford Semele, across the Leam from the Cubbington outliers. Southam Road Borehole [3501 6430], at Radford Semele, proved 2.7 m of brown- and blue-blotched, tough boulder clay, containing flints, chalk, sparsely oolitic Jurassic limestone and small Bunter pebbles, on Wolston Sand and Gravel (Richardson, 1928, p.85). Farther north-east, Offchurch No. 2 [3622 6548] and No. 2a [3623 6538] boreholes proved 7 and 10 m respectively of Oadby till. In the former, the top 1.5 m was red-brown, vaguely laminated in parts, and contained flints but no chalk or Jurassic limestone; although here classified as decalcified Oadby Till, this material may be Upper Wolston Clay.

Several detached patches of boulder clay are preserved due south of Royal Leamington Spa. At Whitnash [322 625] a grey to brown clay contains flint and Bunter quartzite pebbles, together with abundant chalk where unweathered. It also includes red pebbly clay and red-brown silty lithologies like some of those in the Wolston Clay. The larger outcrop near Harbury is somewhat similar.

A number of isolated drift remnants contain tills of variable lithology. Although here described for convenience under the heading of Oadby Till, this correlation is speculative and their age uncertain.

Bishop (1958, pp.270–272) recorded two sections of drift in the disused Harbury quarries. One [391 583] showed 4.6 m of unbedded, grey-brown boulder clay with red and purple streaks, containing Bunter quartzite and Lias limestone pebbles, overlying 2.5 to 3 m of horizontally bedded pink-buff sand and silt, resting on Lias. The other [384 589] showed 3.8 m of cross-bedded, orange and pink sand, overlying 1.5 m of buff, pink-grey and blue-grey silt. In a nearby exposure [382 589] 1.9 m of red, sandy boulder clay, with patches of clayey sand, overlies 4.2 m of dark grey, brown-weathering boulder clay, which rested on 1.9 m of brown silt and sand underlain by 1 m of red and green sandy boulder clay. The boulder clays contain pebbles of Bunter quartzite and Triassic sandstone, siltstone and mudstone, a few of Jurassic limestone (mostly Lias), and coal and flint. The top bed is particularly pebbly and may be Dunsmore Gravel.

There is an even more detached patch at Stockton where disused sand-pits [4359 6411] expose 2 m of blue-grey clay with scattered chalk, flint and Triassic mudstone pebbles, which rest on 0.5 m of red Trias-rich boulder clay. Beneath this is 0.2 m of grey-brown lacustrine-type clay, then 3.0 m of red-brown sand and, at the base of the section, 5 m of red Trias-rich till. The correlation of these lower beds is uncertain. Almost 9 km to the ESE, Thompson (1899, p.82) recorded 3.0 m of red boulder clay overlying up to 1.5 m of sand and gravel on Lower Lias in the Great Central Railway cutting [520 609] north of Catesby Viaduct. The boulder clay is largely of Triassic-derivation and the gravel contains predominantly Bunter quartzite and Liassic ironstone pebbles. These Trias-rich deposits are similar to the till at the base of the section at Stockton.

DUNSMORE GRAVEL

The Dunsmore Gravel, commonly 3–4 m thick and rarely over 6 m, caps much of the plateau between Frankton and Clifton upon Dunsmore. Outliers assigned to it occur as far south as Harbury and as far west as Pinley in south-east Coventry.

It consists of brown, commonly ochreous, poorly sorted, sandy and clayey gravel containing lenses of sand, and is generally particularly clayey in its lower part. The surface layers are especially gravelly, possibly as a result of frost action. The pebbles are mostly of flint (in places up to 50 per cent), Bunter quartzite, Carboniferous sandstone and Jurassic ironstone, with some of Triassic sandstone. Where unweathered, the gravel also includes Jurassic limestone and chalk. Invariably the upper layers are leached and decalcified, with a layer of iron-pan 0.3 m to 0.6 m below the ground; in places a hard 'motherstone' has formed, consisting of gravel strongly cemented by limonite. Derived fragments of this ferruginous rock occur in alluvial fan gravels at Stretton-on-Dunsmore (p.59).

At Clifton upon Dunsmore, the gravel overlies Oadby Till, and it probably represents outwash from the Oadby Till ice (Shotton, 1953). Where seen, its base is erosive and irregular. Broadly, however, depressions in the base correspond with the valleys of the River Avon, the Clifton Brook and the River Leam, showing that the Avon-Leam drainage was functioning when the Dunsmore Gravel was laid down. The long-profile of the surface of the Dunsmore Gravel along the Avon valley appears to grade into that of the 'Fifth Terrace' downstream (Tomlinson, 1925, fig. 2, 1935). This latter has, however, recently been shown to comprise two quite separate terraces (Wyatt, 1982), and the equivalence is with the surface of the upper of these. Temperate fossils recorded from the 'Fifth Terrace' (Shotton, 1977, p.10) and from its equivalent on the Severn (the Bushley Green Terrace of Kennard, in Wills, 1938, p.173), probably all come from the lower terrace. It seems likely that the Dunsmore Gravel predates this temperate episode.

Details

Royal Leamington Spa to Rugby

The outwash fan is preserved continuously from the eastern margin of the district westwards to Frankton. Farther south-west a few minor outliers presumably represent its continuation, as west and north-east of Cubbington [336 684; 350 692] where the deposit is rarely more than 1 m thick, though there was 1.8 m of pebbly sand in Cubbington Borehole.

Burnthurst Farm Borehole [3880 7158], on the south-east of a much larger outlier, proved 1.5 m of red-brown sand, increasingly clayey downwards and containing scattered flint and Bunter quartzite pebbles, on Wolston Clay (Sumbler, 1981). On the main outcrop, boreholes at Blue Boar road-bridge [4534 7195] proved 4 m of yellow-brown sandy clay with lenses of gravel and clayey sand, passing down into brown clayey gravel which rested on Wolston Clay. The nearby railway cutting [459 721] shows 3 m of poorly sorted, brown, sandy and clayey gravel, very clayey in the lowest 0.5 m, with irregular lenses of brown, poorly cemented sand.

Boreholes in Rugby proved, above Wolston Clay: 3.8 m of brown and yellow, gravelly sand with clayey pockets [4873 7297]; 3.1 m of brown clayey and silty sand with fine- to medium-grade chalk and flint pebbles [4990 7520]; 4.6 m of sandy, clayey gravel [5054 7539]; 6.8 m of brown, sandy and flinty gravel [508 745]. In this same general area, Wilson (1870a, 1870b) described the Dunchurch Road Pit [502 744] as showing, above Wolston Clay, more than 4 m of Dunsmore Gravel, consisting of sugary sand with chalk grains, overlain by flint and quartzite gravel. The base of the deposit was uneven and pieces of Wolston Clay were incorporated in the sand.

The Great Central Railway at Rugby cuts through Dunsmore Gravel for a distance of over 2½ km [5150 7543 to 5163 7288]. Thompson (1899) described the gravels as being about 3 m thick with two pockets (?channels) extending downwards for 5 to 6 m to the base of the cutting [5140 7513; 5149 7382]. The gravels were irregularly bedded, coarse and dirty, perhaps rather cleaner and richer red in the southern part of the cutting, and contained pebbles of chalk, flint, quartzite, quartz and gritstone.

In Vicarage Hill Borehole [5244 7598], 1.2 m of red-brown loamy sand, increasingly clayey downwards with abundant flint and Bunter quartzite pebbles, overlay Oadby Till (Sumbler, 1981).
MGS

Outliers to the south

Many small patches of sand and gravel at the top of the drift sequence lie south and west of the main fan, and some are associated with till and laminated clay. Most have been referred to the Dunsmore Gravel, though not all necessarily ever formed part of a single spread. There are several such patches between Radford Semele and Hunningham, for the most part of flinty gravel though some are almost wholly sand. Offchurch No. 2 Borehole proved 3 m of gravel, clayey towards the top, and a temporary excavation [3562 6545] showed 2.4 m of gravel containing bodies of purple-brown ironpan in the upper part. Flints are rare or absent around Offchurch, and an excavation [3761 6554] near Burnt Heath Farm showed 0.75 m of pale grey-brown to buff sand with Bunter quartzite pebbles, overlying 2 m of red orange-brown, clayey sand containing Bunter quartzite and Carboniferous and Triassic sandstone, with irregular lenses and pockets of red-brown boulder clay. The lower sand had a cryoturbated top, and lobes of the upper sand extended down for 0.5 m below the general level of the junction. KA

There are other outlying patches of gravel around Draycote Water. In one, on Hensborough Hill, 0.5 m of red-brown, clayey gravel with flint, Bunter quartzite and sandstone pebbles is exposed in an old pit [4617 6914]. In another on Bunkers Hill [4816 6931], 0.8 m of orange-brown, pebbly, sandy clay and clayey sand is exposed. KA

Near the southern extremity of the district near Harbury, several gravel outliers are scattered along the crest of the Penarth Group scarp. Temporary sections showed 1 m of cryoturbated pebbly sand with a few quartzite boulders [3786 6066] and 1.5 m of horizontally bedded Bunter quartzite gravel with a brown sandy matrix [3788 6227]; the deposits also contain a few flints and Jurassic rocks. JB

About 5 km to the east, Bishop (1958, pp.267–270) described a channel cut into the Lias on the top of Ladbroke Hill [431 595]; it contains 1.2 m of stoneless silts and silty sand, overlain by up to 1.8 m of heterogeneous, reddish brown, sandy, gravelly clay, containing pebbles of Bunter rocks (56 per cent), sandstone (8 per cent) and local limestone (35 per cent), but no flint. Boreholes have since shown that the gravelly clay is overlain by, or passes into, sand and gravel formerly worked for building sand [4312 5949]. This sand and gravel, together with the gravelly clay, are tentatively assigned to the Dunsmore Gravel.

RIVER TERRACE DEPOSITS

Fluviatile deposits form substantial spreads in the valleys of the Avon and its tributaries. All post-date the suite of glacial and fluvio-glacial deposits and are broadly similar in lithology, being composed of gravelly sands with lenses of silts. The pebbles are mainly Bunter quartzites and flints, though Triassic sandstones and Liassic limestones are common along the Leam and Itchen.

Most of the deposits are preserved as flat-topped terraces, and the terrace-flats have been classified according to their levels above the present alluvium. There are four main levels, though there are some deposits that are higher than the highest terrace-flat. The 1st Terrace of the Avon is about 1.5 m above the alluvium near Rugby, and 3.0 to 3.5 m above it downstream from Warwick. The main flat (2 or 2a) of the 2nd Terrace is about 5 m above the alluvium. Below the confluence of the Avon and the Sowe, contiguous deposits rise locally to a less definite flat (2b) slightly higher than the main terrace. The higher terrace-flats are more fragmentary. A few small shelves between the 2nd and 4th terraces have been tentatively assigned to the 3rd Terrace, but there are too few of these to be certain that all lie on the same thalweg. The 4th Terrace lies some 10 to 12 m above the alluvium at Rugby, increasing downstream to over 20 m at Wasperton; for convenience some higher deposits have also been classified as 4th Terrace.

Even within the district this classification is far from satisfactory. In particular it is possible that the 1st and 2nd terraces of the Avon above Warwick correlate with the two benches of the 2nd Terrace (2a and 2b respectively) below that town, and that the terraces of the Sowe and of the Leam above Birdingbury may not agree in their numbering with those of the Avon. Detailed levellings along all the terrace-flats would need to be carried out to resolve the many problems that have arisen, and correlation of the flats with others farther downstream in the Avon catchment can be only tentative.

Classification of the deposits, as opposed to the terrace-flats, is even more uncertain, largely because there is little to show which flats are constructional and which have been cut across pre-existing terrace gravels. The available faunal evidence is, at best, scanty. South of Warwick, *Mammuthus primigenius* and Acheulian-type artifacts have been obtained from gravels assigned to the 4th Terrace at Heathcote [3076 6246] (Playle, 1962) and Barford [c.278 614] (Jack, 1922; Tomlinson, 1935). About 15 km to the north-east alluvial fan gravels [415 738] near Stretton-on-Dunsmore, broadly of the same age, have yielded *Equus caballus*. Both faunas indicate that the gravels were laid down in cold conditions, and this is substantiated by cryoturbation structures within the deposits. In contrast, Tomlinson (1925) has recorded an interglacial fauna held to be Ipswichian from gravels beneath the Avon 3rd Terrace-flat around Evesham, some 40 km downstream from Warwick.

So, too, has Shotton (1929) from a deposit in the valley of the Sherbourne [323 791], a tributary of the Sowe, at Coventry, which he accordingly correlates with Tomlinson's 3rd Terrace. Cold climate mammalian faunas occur in the 2nd Terrace deposits downstream from the district (Tomlinson, 1925), and possibly also at Little Lawford [464 773] (see p.60): they are dominated by *Mammuthus primigenius* and *Coelodonta antiquitatis*. Extensive subarctic insect faunas occur in the 2nd Terrace near Brandon [390 754] and near Evesham, and associated peats have yielded mid-Devensian radiocarbon dates of about 30 000 and 38 000 BP respectively (Coope, 1968; 1962). No fauna is known from the 1st Terrace deposits, but these are presumably late-Devensian to early Flandrian in age.

Interpretation of this faunal evidence demands that a distinction be drawn between terrace-flats and terrace-deposits; only rarely can this distinction be established with certainty. Tomlinson (1925), noted similarities in the faunas from the gravels beneath the Avon 3rd Terrace-flat at Evesham and the 4th Terrace-flat of the Stour near Stratford; she concluded that the terrace gravels were part of a single aggradational episode, that the terrace-flats were erosional, and thus that the gravels beneath the 3rd Terrace were somewhat older that those beneath the 4th Terrace, with the 3rd Terrace-flat being younger than both. Despite the lack of conclusive evidence Edmonds and others (1965) and Williams and Whittaker (1974), in the adjoining Banbury and Stratford upon Avon districts, took the stance that all the terrace-flats were aggradational, and thus that increased height of deposits above the alluvium equated with increased age: the present authors agree with this view, although the survey of the Warwick district has thrown no further light on the matter.

FOURTH TERRACE AND ALLUVIAL FAN GRAVELS

Extensive 4th Terrace-flats occur in the valley of the Avon and the lower reaches of the valley of the Leam—there are smaller patches in the valleys of the Sowe and Finham Brook. The spreads must once have been over 2 km wide at Ryton-on-Dunsmore and over 5 km wide below Warwick. Despite this the underlying deposits are rarely more than 5 m thick, and commonly much thinner.

The 4th Terrace-flat is about 11 m above the modern flood-plain at Rugby, and the deposits lie up to 25 m above the flood-plain at Wasperton. Below Warwick the deposits are rarely built-up to a flat, but underlie gently sloping ground ranging from about 21 to 33 m above the flood-plain, though the gravels are seldom more than 5 m thick. In places gravels extend up the valley sides well above the flat; they may be head, derived contemporaneously from the Dunsmore Gravel, or the remains of fans deposited by streams that once debouched onto the terrace. Isolated patches of flinty gravel at Ashow [313 709], Hill Wooton [303 689] and Old Milverton [303 675] stand several metres above the 4th Terrace-flat, and are separated from it by a rock-bench; for convenience they are here treated as 4th Terrace-deposits, though they may equate with the 5th Terrace gravels in the adjoining Banbury district (Edmonds and others, 1965).

Cryoturbation structures have been recorded within the gravels at several localities, showing that accumulation took place in a cold climate. This is confirmed by the fauna at Heathcote and Barford (see p.58).

Between Wolston and Stretton-on-Dunsmore a set of gravel fans are remnants of the filling of a stream channel that post-dated the deposition of the Dunsmore Gravel and drained southwards to the Leam. Shotton (1953) included them in the Dunsmore Gravel, but they are appreciably lower in level, and the channel thalweg grades southwards into that of the 4th Terrace of the Leam. Another patch of gravel [4433 7615] near Church Lawford lies well above the 4th Terrace-flat, and is probably an analogous fan.

Details

River Avon

Two small patches of red-brown clayey and silty gravel occur just north of Clifton upon Dunsmore [534 769; 5298 7683]. The gravels in the larger patch crop on a slope, that rises for about 8 m and has no obvious flat.

The deposits in Rugby Quarry [4949 7588] comprise 3 m of ochreous, clayey sand, containing pebbles of Bunter quartzite, flint and Carboniferous sandstone. Wilson's (1870a, 1870b) description of the gravels suggest that they have been affected by cryoturbation. An old pit [4481 7736] west of King's Newnham, and above the terrace flat, shows 2 m of orange-red, sandy gravels of flints and Bunter-type pebbles. Some 1.4 m of similar gravel in another old pit [4389 7637] is strongly cryoturbated and contains lenses of sand and large frost-wedges filled with sparsely pebbly, brown sand.

Shotton (1968, pp.390, 391) recorded up to 4.6 m of terrace gravel east of Brandon [411 763]. West of the village the gravels were formerly extensively worked [384 758], and Shotton recorded up to 5.5 m of red clayey gravel, poor in flint, overlain by up to 2.4 m of ochreous flinty gravel. Solifluction festoons occur low down in the flinty gravel, and frost wedges extend down from its top.   MGS.

River Sowe

Two small patches of gravelly sand on the southern side of the Sowe [350 755, 352 754] have flat tops about 14 m above the modern flood-plain. A small patch of flinty gravel in Willenhall [370 769] seems too high above the river to fit the profile of the 4th Terrace-shelf.

Several of the outcrops of gravel north and west of King's Hill [327 746] were regarded by Shotton (1953) as 4th Terrace-deposits but seem too high for this to be so and may be Baginton Sand and Gravel. Only one small patch [3215 7440] is here assigned to the 4th Terrace-deposits; it is conspicuously flinty, and there is a marked feature where it adjoins the Baginton Sand and Gravel.   MGS, RAO

River Leam

There are a few deposits along the main valley, and west of Eathorpe [386 693] the terrace can be divided into two levels, the upper (4b) standing 2 to 3 m above the lower (4a).

Deposits on the interfluve between the Leam and the Itchen include those at the site of the disused Marton railway station [4133 6805], where up to 1 m of medium-grade pebbly gravel with flint, Bunter quartzite, Lias limestone and ironstone, Jurassic fossils and Mercia Mudstone in red loamy sand, is exposed. Farther south, 3.5 m of yellow-brown sand with few pebbles are exposed in an old pit [4146 6724]. Shotton (1953, p.235) has suggested that the deposit marks an old course of the Itchen running between Long

Itchington and Birdingbury; alternatively it may follow a former course of the Leam, joining that river to the Itchen.   KA

Stretton-on-Dunsmore

Alluvial fan gravels probably associated with the 4th Terrace occur between Wolston and Stretton-on-Dunsmore. They are remnants of the filling of a channel which drained southwards into the River Leam. Trial boreholes north of Frog Hall [415 735] showed up to 9 m of gravel in a channel cut into Wolston Clay. The gravel, which is commonly ochreous near the surface, consists of orange-brown sand containing pebbles of flint and Bunter quartzite together with a few limonitic fragments apparently derived from the Dunsmore Gravel iron pan. The fan gravels are better sorted and less clayey than the Dunsmore Gravel and have been worked at a number of places. One small gravel pit [4052 7191] showed limestone pebbles apparently derived from nearby outcrops of the Langport Member. A pit at Frog Hall showed 6 m of gravel and has yielded a metatarsal of *Equus caballus* (Shotton, 1953 p.226).   MGS

THIRD TERRACE DEPOSITS

Few 3rd Terrace deposits have been mapped, and of these some may belong to the 4th Terrace and some may be head. Shotton (1953, Fig. 9) has assigned gravels at Newbold on Avon [485 764] and at Lower Farm, Brandon [375 757] to the 3rd Terrace. No terrace deposits were found at the first of these localities during the present survey, and at the second the deposits, about 5 to 7 m above the floodplain, seem more likely to be 2nd Terrace. At Old Milverton [299 675] a deposit of flinty gravel capped by a poor bench about 11 to 12 m above the floodplain, is tentatively regarded as 3rd Terrace. At Castle Park [287 638; 286 636] deposits of red sandy gravel, contiguous with 4th Terrace-gravels, are benched at about 11 to 15 m above the floodplain, several metres below the adjacent 4th Terrace-flat and there are two analogous gravel benches north-west of Wasperton [260 598; 260 595].

A deposit of flinty gravel along the Sowe at Whitley [356 767], is about 9 to 12 m above the floodplain, and lies midway between the profiles of the 2nd and 4th Terrace-flats. Up to 3.7 m of sand and gravel [3565 7668], with Bunter quartzite, flint and Triassic skerry pebbles, have been recorded resting on Bromsgrove Sandstone.

On the Leam three outcrops of flinty gravel between Hunningham and Weston under Wetherley [372 678; 375 678; 369 688], lie about 13 to 14 m above the modern floodplain, and so have been assigned to the 3rd Terrace: the first of these was included in the 4th Terrace by Shotton (1953, fig. 9).

SECOND TERRACE DEPOSITS

Deposits of the 2nd Terrace are widespread in the modern winding valleys; the terrace-flat lies 5 to 6 m above the floodplain. In places the flat is poorly defined and gravels extend up the valley sides from it. The deposits are generally more than 5 m thick and thus commonly continue below the level of the 1st Terrace-flat with no separating rock step.

Cold-climate mammalian fossils have been recorded outside the district (p.58), and some at least of the following fauna from near Little Lawford [464 773], originally described by Buckland (1823), probably came from gravels of the 2nd Terrace: *Mammuthus primigenius*, *Coelodonta antiquitatis*, *Bos primigenius*, *Equus caballus* and *Crocuta crocuta* (Sumbler, 1983a, b). Mammalian fossils, now in Warwick museum, found at Charlecote [266 576], just beyond the southern margin of the district, include *M. primigenius*, *C. antiquitatis*, *Rangifer tarandus* and *Equus* cf. *przewalskii* with *Felis leo* recorded additionally by Shotton (1976b); *M. primigenius* was also found in 2nd Terrace gravels at Old Warwick Road, Royal Leamington Spa [314 652]. The deposits have also yielded flora near Brandon [390 754] (Kelly, 1968), and a fauna comprising *Dicrostonyx henseli*, molluscs (Shotton, 1968), abundant arctic insects (mainly coleoptera), and some arachnids (Coope, 1968). The fossil evidence points to an environment comparable to the tundras of present-day northern Europe. Three samples of peat from this locality have yielded radiocarbon dates of around 30 000 years BP (Shotton, 1968, p.388).

Details

River Avon

At Newbold on Avon the 2nd Terrace of the Avon is recorded on BGS field maps of 1914 as showing 3 m of brown sand and gravel, with flint, Bunter quartzite and Lias pebbles, resting on Blue Lias [4929 7698]. A claypit [4656 7732] at Little Lawford exposes up to 2 m of brown sand and gravel, with Bunter quartzite, flint and sandstone pebbles; both the gravel and the underlying Lower Lias clay show cryoturbation involutions.

Shotton (1968, pp.389, 390) described dark grey silt and peat south-west of Brandon [390 754]; this material yielded the flora and fauna mentioned above. Similar deposits occur in a gravel pit north of Ryton-on-Dunsmore, where the section is: [3824 7494 to 3815 7500]:

	Thickness m
Gravel; brown, ferruginous, weakly cemented, medium-grained sand matrix with Bunter quartzite and flint pebbles	1.3
Sand, orange-brown, medium- to coarse-grained; increasingly grey downwards with silt and pebbly lenses; clayey near base	1.7
Clay; pale blue-grey, smooth dark and peaty in lower part, with much finely divided plant debris; pebbly near base	0.4
Gravel, brown, ferruginous with large cobbles in lower part	1.5
Mercia Mudstone	0.6

Below the confluence with the Sowe, the 2nd Terrace-flat is commonly divisible into two (2a and 2b). Terrace 2a forms the main flat, and is usually around 5 m above the floodplain and commonly several hundred metres wide. Terrace 2b has a poor flat and the deposits reach up to about 10 m above the floodplain. Excavations in the Woodloes Housing Estate [281 660], Warwick, showed sand, sandy clay and gravel, with pebbles of flint and Bunter quartzite. These deposits are 5.9 m thick in Emscote Mills No. 2 Borehole [2928 6570]. MGS

River Sowe

The 2nd Terrace of the Sowe west of Tollbar End is cut by a ditch [3521 7561] showing 3 m of loamy gravel containing much red and grey clay; a nearby borehole [3530 7558] proved 4.6 m of compact sand and gravel with large cobbles, resting on Mercia Mudstone.

Excavations at Finham Sewage Works [3375 7407] exposed 1 to 2 m of flinty gravel with sandy and clayey bands. At Stoneleigh [3311 7275], excavations revealed 0.6 m of bedded orange-brown sand and red-brown silty sand, overlying at least 0.8 m of bedded sand and gravel with angular flints and scattered bone fragments. MGS, RAO

Rivers Leam and Itchen

At Royal Leamington Spa the gravels are generally about 4 m thick, but one borehole proved 6.3 m (unbottomed). The main terrace-flat lies about 6 m above the modern floodplain, but gravels commonly extend for 3 m or more above it. Between Long Itchington and Hunningham the terrace-flats are commonly at rather different heights on opposite sides of the river, suggesting that some are erosional trims. KA

FIRST TERRACE DEPOSITS

The 1st Terrace flat adjoins the floodplains of the Avon and Leam throughout most of their courses, forming a bench (1 or 1b) about 1.5 to 3.5 m above the alluvium. In a number of places there is a lower flat (1a). The deposits consist of loamy sands and gravels, commonly clayey near the surface and rarely more than 3 to 4 m thick; they are generally of finer grade than those of the older terraces.

Details

River Avon

The 1st Terrace of the Avon in Rugby was penetrated by boreholes [5020 7624] which proved 3.5 to 4.3 m of clayey, brown sand with scattered pebbles, coarse and gravelly in the basal 0.8 m. A riverbank section [3658 7510] south of Tollbar End shows 1.3 m of sandy gravel with pebbles of flint and Bunter quartzite, resting on Mercia Mudstone; the basal 0.4 m of gravel has been cemented by blue-black limonite. Boreholes west of Bubbenhall proved up to 4 m of brown, sandy clayey gravel [3525 7238]. The river-bank [3478 7245] to the west exposes 1.8 m of very flinty gravel; abundant fragments of Enville Group Sandstone near the base of the gravel have been derived from the underlying rock.

River Sowe

West of Willenhall the 1st Terrace of the Sowe has been penetrated by boreholes which proved up to 3.7 m gravels. The river bank [3412 7441] at Baginton shows 0.7 m of sandy alluvium on 0.3 m of coarse, brown sand on 0.6 m of bedded sand and gravel. MGS, RAO.

River Leam

At Eathorpe exposures in the terrace-deposits include 1.3 m of red-brown gravel with clayey pockets and pebbles of flint, Bunter quartzite, and sandstone from the Mercia Mudstone [3912 6927]. A ditch [3988 6575] along the Itchen exposes 1.5 m of gravel with pebbles of flint, Bunter quartzite, and limestone from the Lias and the Langport Member.

The fillings of an abandoned river channel at Marton Moor [409 679] are probably coeval with the 1st Terrace gravels. The deposits are grey clayey loams, pale green-grey at depth. KA

ALLUVIUM

Alluvial deposits occur in all but the smallest valleys. Their composition reflects local sources, both solid and drift, and even in the largest valleys they rarely exceed 6 m in thickness.

Details

The alluvium of the Avon has been proved by boreholes at Rugby to comprise silty clay and loam on sand and gravel, and to be more than 3.4 m thick. A trench in the floodplain of the River Swift [5021 7737 to 5032 7733] exposed 0.3 to 0.7 m of clayey loam on 2 to 3 m of sandy gravel resting on Lower Lias. The gravel was mainly composed of flint and Bunter quartzite pebbles, with some large cobbles of flint, Carboniferous sandstone and Leicestershire diorite. Trial boreholes for gravel in the floodplain north-east of Ryton-on-Dunsmore showed about 2.4 m of alluvium; a borehole near the centre of the floodplain [3979 7581] proved 1.5 m of grey-brown clay, on 4.6 m of clayey sand and gravel resting on Mercia Mudstone. A river-bank section [3919 7462] near Ryton-on-Dunsmore showed 0.5 m of brown loam overlying 1.5 m of brown clay with scattered pebbles on coarse gravel.

The alluvium of the Leam at Royal Leamington Spa has been proved by boreholes to comprise about 3.4 to 5.4 m of clay on up to 1.5 m of gravel. Leamington Bypass No. 27 Borehole [3077 6539] penetrated 7.0 m of clay on 0.6 m of gravel.

Dury (1952) described the alluvium of the Itchen at Depper's Bridge [400 593] as comprising plastic clay underlain by silt, together up to 3.6 m thick; he concluded that it occupied a channel formed by a stream much larger than the present river.   MGS, KA.

PEAT

Thin peat is common in local marshy areas of river floodplains, but mappable deposits exist at only a few places; most are less than 1 m thick. Lenses of peat, formed from rafts of plant debris, occur locally within alluvial deposits. Peat occupies the ground adjacent to Clifton Brook and one of its tributaries at Hillmorton [535 747]. It averages about 1 m in thickness but in places exceeds 2 m. The deposit occurs where the permeable Hillmorton Sands fill a channel bounded by clays. Its outcrop was once a marsh and the peat, together with the underlying quicksand, caused considerable difficulties during construction of the canal and railway (Wilson, 1870a, p.32, 1870b, p.200, 1875, pp.10, 11).   MGS

HEAD

Solifluction deposits are widespread on and below hill slopes, their lithologies reflecting the local geology. Only the thickest and most extensive spreads have been separately distinguished on the published map.

Deposits at and to the east and south of Radford Semele [346 647] are largely composed of gravel derived from the drift: a riverbank exposure [3533 6493] is in 1.5 m of poorly sorted clayey unbedded Bunter-quartzite and flint-pebble gravel. Several areas of red-brown sandy to silty clays border the alluvium between Radford Semele and Bascote Locks [380 638]; the deposit rarely exceeds 1.5 m in thickness.

A section below the Penarth Group scarp at Bascote [3970 6323] shows 0.7 m of mainly grey-green clay, resting on Mercia Mudstone. The basal 0.1 m contains much red clay, orange-brown to grey sand, and Langport limestone.   KA

LANDSLIPS

Landslips are characteristic of the Penarth Group scarp, where relatively resistant rock overlies clays. Most are either multiple small-scale rotational slides or mudflows and were probably lubricated by water seeping from the Langport Member. There are several fresh slip-scars on the steeper slopes showing that some movement still occurs. Of the few slips that have affected Liassic strata, the most extensive are in the Middle Lias around Napton on the Hill. Small slips have also occurred in the Wolston Clay at Thurlaston.

Details

On the Penarth Group escarpment below Ranch House, Harbury [352 597], successive small rotational slipping has occurred, and has produced a number of parallel ridges. At Chesterton Green [369 585], the toe of the slipped area is covered by secondary mudflows. South of Lower Westfields Farm, Harbury [361 605], a fresh scar, 0.5 m high and a few tens of metres long, lay about halfway down the face of the escarpment in 1976; it is reported to have formed over the preceding one to two years.

On the lower slopes of the scarp at Ufton, up to 3.9 m of slipped material overlie Tea Green Marl [377 623]. The slipped debris is made almost entirely of Cotham and Westbury mudstones with some Langport limestone and drift pebbles from the crest of the scarp. On the scarp south of Birdingbury, there is a prominent backscar with hummocky slip debris below [422 666]. Fresh slip scars up to 0.5 m high occur nearby [425 666], and the road along the crest of the scarp between 0.5 and 1.3 km south of Birdingbury is very uneven, suggesting that minor movements are still taking place.

East of Marton [421 692], the Langport Member and the Lower Lias are involved in the slipping. North-east of Windmill Hill [412 704] slipped material underlies the alluvium.

Away from the Penarth Group escarpment there are slips along the Middle Lias outcrops around Napton on the Hill [462 614]. A bed of sandstone crops at the top of the slip scar, and gives rise to considerable seepage which probably helped to initiate the slips. Around Beacon Hill, about 2.5 km to the east [490 610], springs rising from the base of the sandy and silty Middle Lias, may similarly have triggered slips in the underlying clays. Small landslips occur immediately east of Barby Wood Farm [524 705] and 1 km farther east [533 705]: they probably result from weakening of the soft silty clays of the *ibex* Zone by seepage from the '85-Marker' which crops out immediately above the slips.   KA, RAO, MGS

CHAPTER 8

Structure

NEAR-SURFACE STRUCTURES

The Warwick district can be divided into three structural units, each elongated in a generally NNW–SSE direction. At different times in the past these units have behaved independently of one another, producing a complex history of basin inversions.

In the Triassic and Jurassic rocks the bounding structures of these units are vague, though the broad effects of movements along them are clear. The eastern unit encompasses the south-eastern extension of the western part of a shallow Triassic basin centred beneath Hinckley, and here termed the Hinckley Basin. Within the district, contours on the base of the Triassic show little indication of its western margin, but they progressively swing in direction from SW–NE to SE–NW across the Coventry district to the north where the basin margin is sharper (British Geological Survey, 1985): possibly the Princethorpe Fault is related to the basin margin at depth. The central unit extends southwards from Tamworth through Coventry to the southern margin of the district; it is here called the Coventry Horst. Triassic outcrops are largely confined to the area south of the Avon, but the projected height of the base of the Triassic from Kenilworth northwards must have been well above its level in the units to the east and west. The western unit lies west of a broad zone of faulting, probably encompassing the Warwick Fault and the Western Boundary Fault in the adjacent Redditch district, and forms the eastern limit of another, and deeper, Triassic basin, here referred to as the Knowle Basin.

The structural relationship within these three areas is quite different in the pre-Triassic rocks. Beneath the Hinckley Basin and its south-eastern extension, Triassic strata rest directly on Cambrian or older rocks, a situation that continues as far north-west as Market Bosworth in Leicestershire (Worssam and Old, in press) where Carboniferous strata incrop against the Trias and dip to the east. The implication is that a west-facing asymmetrical pre-Triassic anticline underlies the western part of the basin, that the once continuous Carboniferous cover has been removed along its axis, and that the post-Triassic basin is a pre-Trias horst. The reverse is true over the Coventry Horst. Here thick Carboniferous and ?Permian rocks have an extensive outcrop and continue beneath relatively thin Triassic cover to form the concealed South Warwickshire Coalfield. Their base lies well below the level of the Cambrian rocks beneath the western part of the Hinckley Basin, and so the structures that separate the two units probably have a different sense of movement in the Carboniferous and older rocks than they do in the Triassic cover, so that the horst becomes a graben in the older rocks. Information is scanty along the west of the horst, and it is uncertain whether the eastern limit of the Knowle Basin lies along the Western Boundary Fault or along the branches of it that cross the Warwick district. It seems possible that these faults have the same sense of move-

ment in the Trias, and that the Carboniferous rocks preserved to the west of the bounding faults do not continue westwards to the axis of the Triassic basin.

The preserved sequences of Carboniferous and Cambrian rocks give no indication of any of the sedimentological changes that might be expected were these structural units in existence during their deposition. It seems likely, therefore, that the controlling structures were initiated during the latest Carboniferous or early Permian. The reversal is likely to have become operative during the early part of the Trias, to have continued more weakly during the later Trias, and to have had only slight effects on the Jurassic sequence.

Within the district, structural evidence comes largely from the Coventry Horst, where there has been extensive exploration for coal. The following account is, consequently, heavily biased towards this area.

Hinckley Basin

Very little information exists within that part of the district that is likely to fall within the Hinckley Basin. A steep late-Permian monocline lies along the edge of the Coventry Horst and is mentioned below (p.64), but the surface evidence in later rocks is masked by Quaternary deposits as far north-east as Dunchurch, while the Lower Lias outcrops still farther south-east yield little firm data. The only proving that may lie east of the horst is Home Farm Borehole. Here Triassic rocks rest directly on the Cambrian, and the surface of the Cambrian rocks is very much higher than it is at Burnthurst Borehole, the next nearest to the west. This suggests that Home Farm Borehole is within the (inverted) basin, but the evidence is not wholly conclusive for it may lie on the monocline along the margin of the horst. If it is within the basin, then the fact that the highest Cambrian rocks in the hole belong to the *C. tenellus* Zone as they do on the adjacent horst means that there has been no movement of the Cambrian basement along the basin margin before the early Permian, and that the horst and basin structures have no earlier history. Moreover, although Carboniferous rocks are absent beneath the axis and western part of the basin, the Westphalian sequences immediately to the west at Binley and to the north-east towards Coalville show no approach to the marginal conditions that would be expected were these major structures active during Carboniferous sedimentation. The gentle flexures in the Triassic rocks suggest that the sagging that took place after the formation of the early Permian anticline beneath the basin was, in general, not controlled by active renewal of movement along older fault-lines, though northwards at Nuneaton the basin margin is sharper and appears to have a longer history than it does within the Warwick district (Taylor and Rushton, 1971).

Coventry Horst

Within this belt there appear to be four distinct episodes of movement, though it is not always possible to isolate the effects of each.

PRE-CARBONIFEROUS MOVEMENTS

The earliest known movements affect the Tremadoc strata, and are pre-Carboniferous in age. Their effect can be deduced from the location of zonally distinctive faunas collected from the lowest cores in the deeper NCB boreholes. The simplest explanation of the disposition of these faunas is that the rocks are affected by an open syncline that trends north-eastwards across the district roughly from Whitnash to Eathorpe, with a complementary parallel anticline beneath Napton on the Hill (see Figure 2); furthermore, in the Coventry district to the north, there is a suggestion of a NNW–SSE syncline trending towards Westwood Heath [285 767]. The data are not sufficient, however, to enable the exact trend and location of these axes to be determined. Many of the recorded dips and minor structures are at variance with this broad interpretation, and there may well be other smaller folds and associated faulting.

INTRA-CARBONIFEROUS MOVEMENTS

Gentle intra-Carboniferous movements are responsible for the unconformity at the base of the Halesowen Formation. Though small in magnitude, the movements had profound economic consequences for they limit the south-eastward extension of the earlier Westphalian strata with their valuable coals. These terminate against the sub-Halesowen unconformity along a line running approximately north-north-eastwards from Ladbrooke (see 1:50 000 published sheet) though, viewed regionally the movements involve gentle uplift along the west–east ridge of St George's Land.

PERMIAN AND TRIASSIC MOVEMENTS

Following deposition of the Enville Group, gentle folding produced a shallow syncline that plunges gently southwards (Figure 19). The syncline appears to flatten northwards from Kenilworth, and is responsible for the major disposition of the outcrops between Kenilworth and Coventry. The Ashow

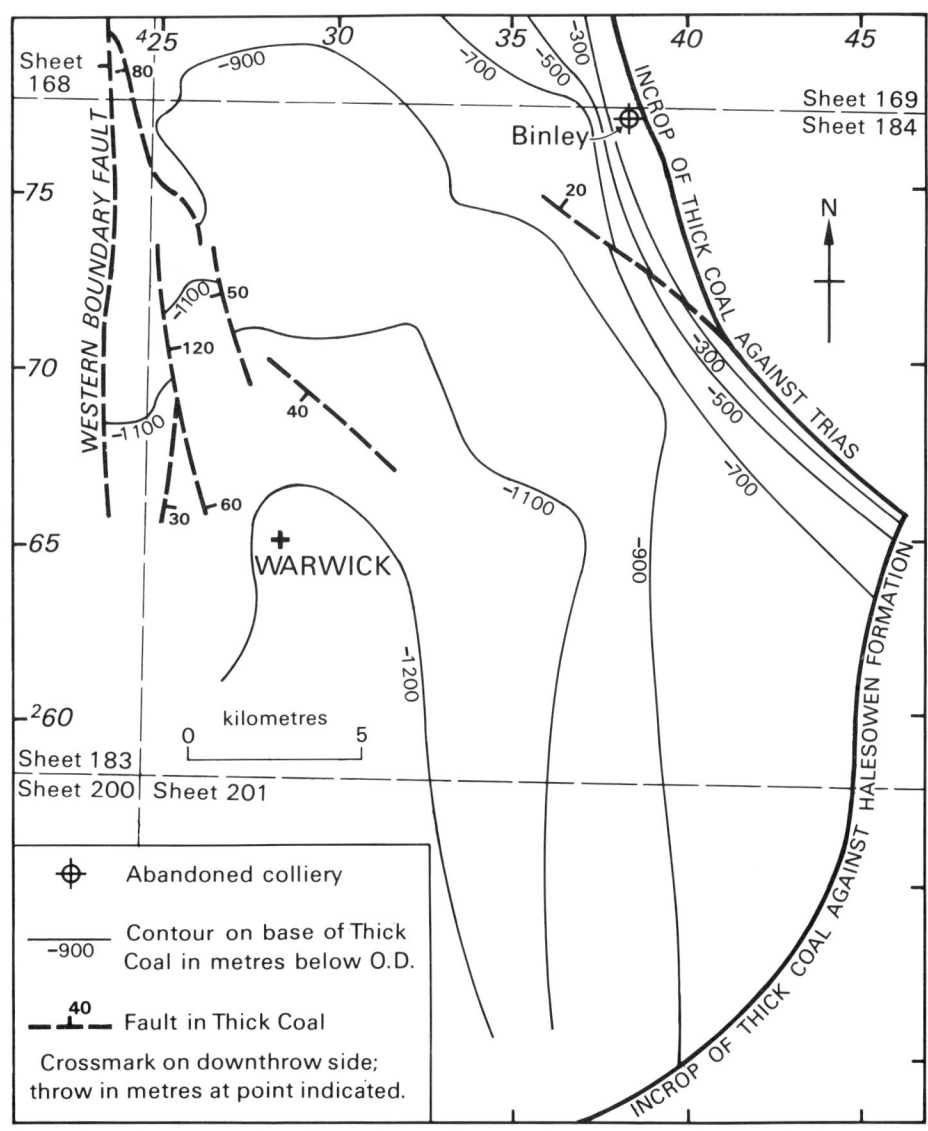

Figure 19   Structure contours on base of Thick Coal (data from National Coal Board)

Formation is preserved on both limbs of the fold at, and to the east of Kenilworth. The older formations strike broadly west–east to the north of Kenilworth on the western limb, but swing almost through a right-angle on the eastern limb beneath the Triassic cover.

The folding was completed before the Bromsgrove Sandstone was deposited because, on the eastern limb of the fold, this formation cuts across outcrops of the earlier rocks; it rests on the Ashow Formation at Warwick, the Kenilworth Sandstone near Stoneleigh, the Tile Hill Mudstone at Baginton, and the Coventry Sandstone east of Whitley Common. The trend of the main fold is sufficiently different from that of the deduced folds in the Tremadoc to suggest that it does not represent renewal of folding along pre-Carboniferous axes.

Although the folding is generally extremely gentle, it is locally steep in a narrow belt extending north-north-westwards from Birdingbury to Binley Colliery, where the westwards dip becomes progressively steeper eastwards. This belt continues northwards to Nuneaton where it is obviously related to the eastern margin of the Coventry horst. Within the Binley workings a north–south monocline, parallel to and presumably associated with this belt of steep dips, produces low westward dips. A number of minor en-échelon folds superimposed on the monocline have dips of 10° to 20° on their flanks, increasing locally to 70° (see Figure 3). In the west there appears to be a west-facing monocline to the eastern side of the Warwick Fault, between Burton Green and Kenilworth. It is discussed further on pp.65, 67.

Many minor faults with north-easterly trends affect the outcrop of the Enville Group. None are large and they may be related to the early Permian folding, though one affects the Bromsgrove Sandstone east of Stareton.

POST-JURASSIC MOVEMENTS

The main effect of post-Jurassic movements has been to impose a gentle south-eastwards dip on the Triassic and Jurassic rocks (Figure 20). There is an abrupt change of strike along the Warwick Fault as the edge of the Knowle Basin is approached, and a much more gradual one west of Rugby over the edge of the Hinckley Basin, though the effects of the latter are more intense northwards in the Coventry district.

Many of the faults in the west are almost certainly a result of the proximity of the margin of the Knowle Basin. They include the Whitnash Fault, the Warwick Fault, and their branches; there is some evidence to suggest that there is thickening westwards in the Triassic across these, implying that movement was active during Triassic sedimentation. The Whitnash Fault, with a downthrow to the west of about 25 m east of Warwick, continues to the southern limit of the district, though its throw here is quite small. The Warwick Fault has a downthrow to the west of about 100 m near Warwick. At surface to the south it passes into the outcrop of the Mercia Mudstone and its continuation is uncertain. While most of this set of faults have westward downthrows, a few of their branches have small throws in an opposing direction. Within the district there is no analogous set along the western margin of the Hinckley Basin. It is possible, however, that the Princethorpe Fault is associated with basin subsidence to the east, for it lies in about the right position

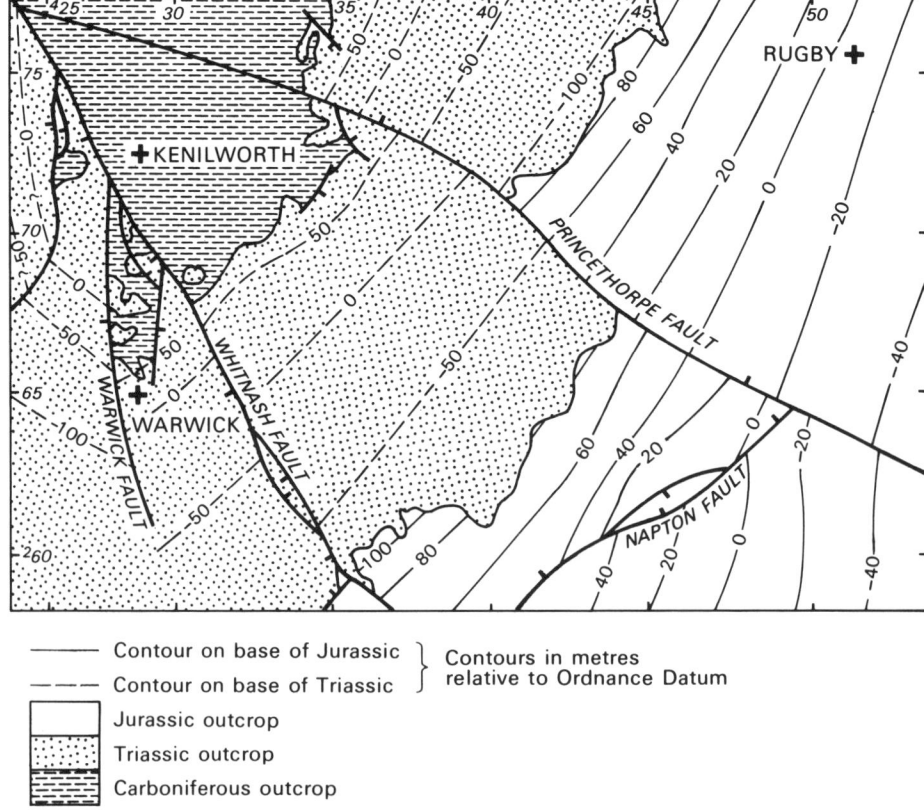

Contour on base of Jurassic ⎱ Contours in metres
Contour on base of Triassic ⎰ relative to Ordnance Datum

Jurassic outcrop
Triassic outcrop
Carboniferous outcrop

Figure 20 Structure contours on the base of the Jurassic and Triassic

south-east of Princethorpe and there is some evidence that it was active when the Arden Sandstone was being laid down. It has a downthrow to the east of about 80 m north-east of Bubbenhall, but this steadily decreases south-eastwards to 40 m near Broadwell and to only a few metres south-east of Flecknoe.

The Napton Fault, and its complementary branch to the north of Napton on the Hill, is unique within the district. The presence of the structures is apparent from the preservation of the Middle Lias at Napton on the Hill well to the north-west of, and at a lower level than, its main outcrop around Hellidon; the Middle Lias outcrop is here within about 1 km of that of the top of the Blue Lias and appears to rest on an anomalously low horizon in the Lower Lias. There are associated minor NE-trending folds in the neighbouring tract to the east. There is no sign of these faults in the Coal Measures and the main fault is presumably a listric one flattening in dip to the north-west, while the branch to its north terminates downwards against it. There is nothing to establish the cause of this composite fracture. Its trend, however, broadly parallels the outcrop of the Lower and Middle Coal Measures against the Halesowen Formation which results from intra-Carboniferous measurements, and it is possible that slight post-Jurassic renewal of uplift along an intra-Carboniferous axis on St George's Land to the south has led to this surface listric faulting.

Knowle Basin

The eastern margin of this basin is marked by several substantial *en-échelon* faults, the largest of which is the Western Boundary Fault in the Birmingham and Redditch districts to the north-west and west (see 1:50 000 published sheet). The overall effect of these structures is to produce a post-Triassic downthrow to the west of at least 200 m so that no coal exploration has gone on along the eastern part of the basin, though the Thick Coal continues to the Western Boundary Fault with little decrease in thickness (Figure 7).

Surface information is meagre. Immediately east of the Warwick Fault there is a sharp anticline, so that dips north-west of Kenilworth are towards the fault. To the west the Bromsgrove Sandstone dips south-west at a gentle angle. Farther west there are no reliable dips in the Mercia Mudstone though the outcrops of skerry bands around Budbrooke point to a continuation of this south-west dip.

Meer End, Wedgnock and Fernhill boreholes, all of which lie west of the Warwick Fault, entered Enville Group beneath the Trias, suggesting that movement along the faults marking the eastern margin of the basin are in the same sense in the Triassic and Carboniferous. This immediately suggests an analogy with the eastern boundary of the Hinckley Basin, where Carboniferous rocks are preserved beneath the Trias along part of the basin eastwards of the pre-Triassic anticline along its axis (see 1:50 000 published sheet 155). It is at least possible that the Knowle Basin behaves in a similar fashion.   RAO

GEOPHYSICAL EVIDENCE FOR DEEPER STRUCTURES

Gravity and magnetic coverage of the district provide further indications of the structural pattern at depth. The differences in density that exist between the proved Triassic, Carboniferous and Cambrian strata give rise to gravity anomalies as their relative thicknesses vary; these sediments are, however, essentially non-magnetic, and the magnetic anomalies derive only from the underlying Precambrian basement and any other younger intrusive or volcanic rocks that may be present. The structural units described earlier in this chapter can thus be delineated more clearly by the Bouguer anomalies (see Figure 21) than by the aeromagnetic anomalies (Figure 22). The main features of the structure of the Warwick district as discussed below are illustrated in Figures 23a and b by profiles of the gravity and aeromagnetic data. The nature of these data is detailed in Appendix 4.

The Bouguer anomaly gradient in the north-eastern extremity of the district is taken to indicate thickening of the Trias into the Hinckley Basin. The axis of the basin probably lies close to the anomaly low outside the district, some 5–10 km beyond Rugby, although this is almost certainly not due entirely to Mesozoic infill. Between this low and another developed near Offchurch there is a broad NW-trending anomaly ridge which appears to be the continuation of a feature culminating near Tamworth at the northern end of the Warwickshire Coalfield. It is intrepreted as a structural high along which the nature of the basement rocks must change: at Tamworth there is a prominent magnetic high; in the Warwick district, the Bouguer anomaly values are 9 mGal lower and the magnetic anomaly is absent, despite the shallow depth to Cambrian strata at Home Farm Borehole. The position of this ridge is consistent with its being the expression of a pre-Triassic anticline underlying or delimiting the western part of the Hinckley Basin. A distinct change may occur along the ridge where the anomaly maximum is displaced near Rugby. This zone is aligned with the supposed SW–NE synclinal axis in the Cambrian through Eathorpe (Figure 2) and the parallel contour trend through Southam towards Rugby; there is also evidence of west–east basement structural trends in both the gravity and magnetic anomaly contours.

The eastern margin of the Coventry Horst is expressed in the Bouguer anomaly gradient that enters the district from the north, passing between Ryton-on-Dunsmore and Stretton-on-Dunsmore before taking on a south-eastward course to Upper Catesby. The steepest gradients as defined on a detailed traverse between Wolston and Cubbington Heath coincide with the fault mapped in the Trias east of Ryton-on-Dunsmore, although they are attributed mainly to steep dips rather than to faulting in the Carboniferous strata. Farther south-east the gradient is offset progressively eastwards to give an echelon form to the margin of the horst. The Princethorpe Fault, which follows the same general direction, here throws the Trias down to the north-east, contrary to the sense of the gradient. Beyond Princethorpe this fault swings round south of Coventry along a secondary trend in the contours, related only in part to the more gradual northward rise in the floor of the Carboniferous. Aeromagnetic values (Figure 22) increase towards the west across the edge of the horst though the con-

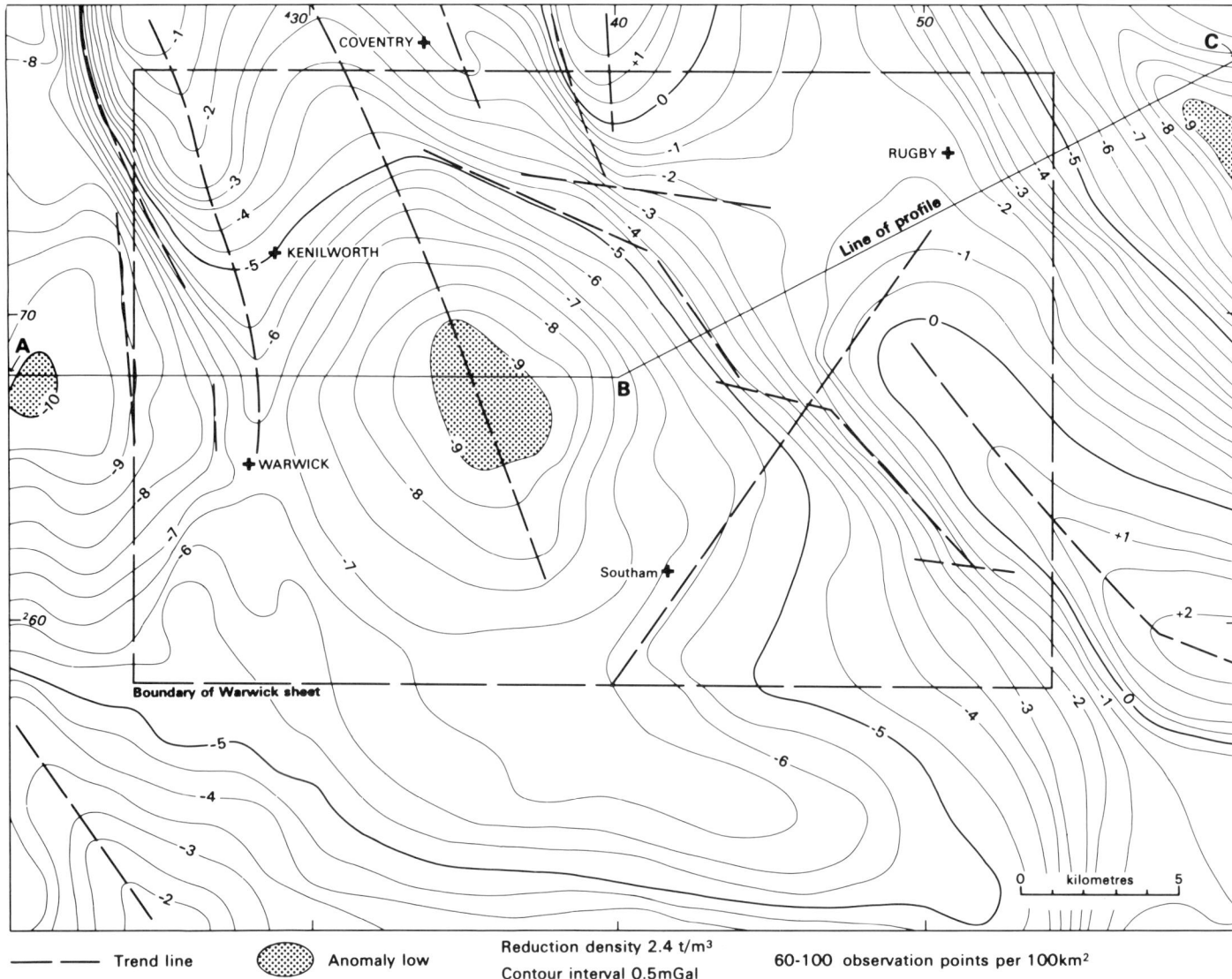

Figure 21 Bouguer anomaly map

tour alignments do not always coincide with the gravity data. The magnetic anomaly gradient may define relief on the Precambrian surface, or more probably a lateral change in its character; in either case it is linked causally to the structural boundary at higher levels and indicates that strongly magnetic rocks rise to a level of about 3000 m below the ground to the west.

The Coventry Horst is dominated by a circular gravity anomaly low centred just west of Offchurch. Triassic rocks can contribute little to this anomaly, and Dunham and Poole (1974) attributed it to the presence of low-density sediments of the type found in Steeple Aston Borehole (Poole, 1977). Such rocks are, however, unknown in the district and alternative explanations are required. The change in gravity value from the ridge in the east to the low can be accounted for solely in terms of the thickening of the Carboniferous sequence, but the increase in anomaly to the north is greater than that predicted on this basis and the minimum does not coincide with the deepest part of the coalfield. Borehole

evidence from farther south (Poole and others, 1968) also demonstrates that the relationships can be more complex, depending on the nature of older formations. Acidic rocks occur beneath parts of the East Midlands (Hains and Horton, 1975; Le Bas, 1972) and the presence of granitic plutons has also been suggested (Evans and Maroof, 1976). Certainly, the association of the Bouguer anomaly low with an area of slack magnetic gradient is consistent with there being rocks of acid–intermediate composition within the basement. The magnetic anomaly contours show west–east trends across the horst, with a high near Kenilworth and a trough extending through Warwick that might both represent flexures in the basement. North–south trends are much weaker, but there is an interesting correlation between the more intense folding observed at Binley Colliery and a 'promontary' in the magnetic relief: the sharp curvatures and steeper gradients indicate a shallower depth to magnetic rock in this area, while meanders in the Bouguer anomaly contours reflect the local structure in the Carboniferous. The axis of the gravity low trending NNW from Offchurch is displaced east of the Carboniferous syncline through Westwood Heath. Both this and the residual increase in anomaly within the horst northwards from Stoneleigh could be most simply explained by corresponding changes in thickness of Cambrian sediments if it were not for the inverse relation to the magnetic anomaly variations. The density contrast at the base of the Cambrian to the south may be slightly negative rather than positive. Taking the basement of the Welsh Borderland as an analogy, volcanic (Uriconian) rocks showing some bulk magnetisation beneath Offchurch could give way northwards to denser (Longmyndian) meta-sediments: evidence from the Nuneaton inlier at the edge of the horst suggests that a Charnian equivalent is present.

The western margin of the Coventry Horst is also marked by a strong Bouguer anomaly high which can be traced to the Cambrian inlier at Dosthill about 25 km north of where it enters the district near Burton Green. Its amplitude decreases continuously to the south, and the feature is lost beyond Warwick in much the same way as is the surface ex-

pression of the Warwick Fault which it closely parallels. The ridge is associated, locally at least, with a sharp, post-Carboniferous anticline which lies on the eastern side of the fault. A magnetic anomaly high west of Kenilworth correlates with the fault, but there is only slight indication of its trend. There is no known thinning of the Carboniferous sequence to explain the increase in the gravity anomaly: a greater number of lamprophyre dykes in the Cambrian could contribute to the higher anomaly values but their origin is thought to lie mainly within a relatively narrow zone of the deeper basement controlled by major faulting.

The Western Boundary Fault is mapped as passing just outside the district; west of Burton Green this line corresponds with the belt of steep gravity anomaly gradient flanking the marginal ridge, as would be expected if the near-surface Triassic sediments thicken rapidly across it. Maximum gradients occur about 2 km west of the Warwick Fault where they enter the district, with the contours heading SSE rather than to the south. Approaching Warwick the main gradient slackens rapidly before the contours swing to

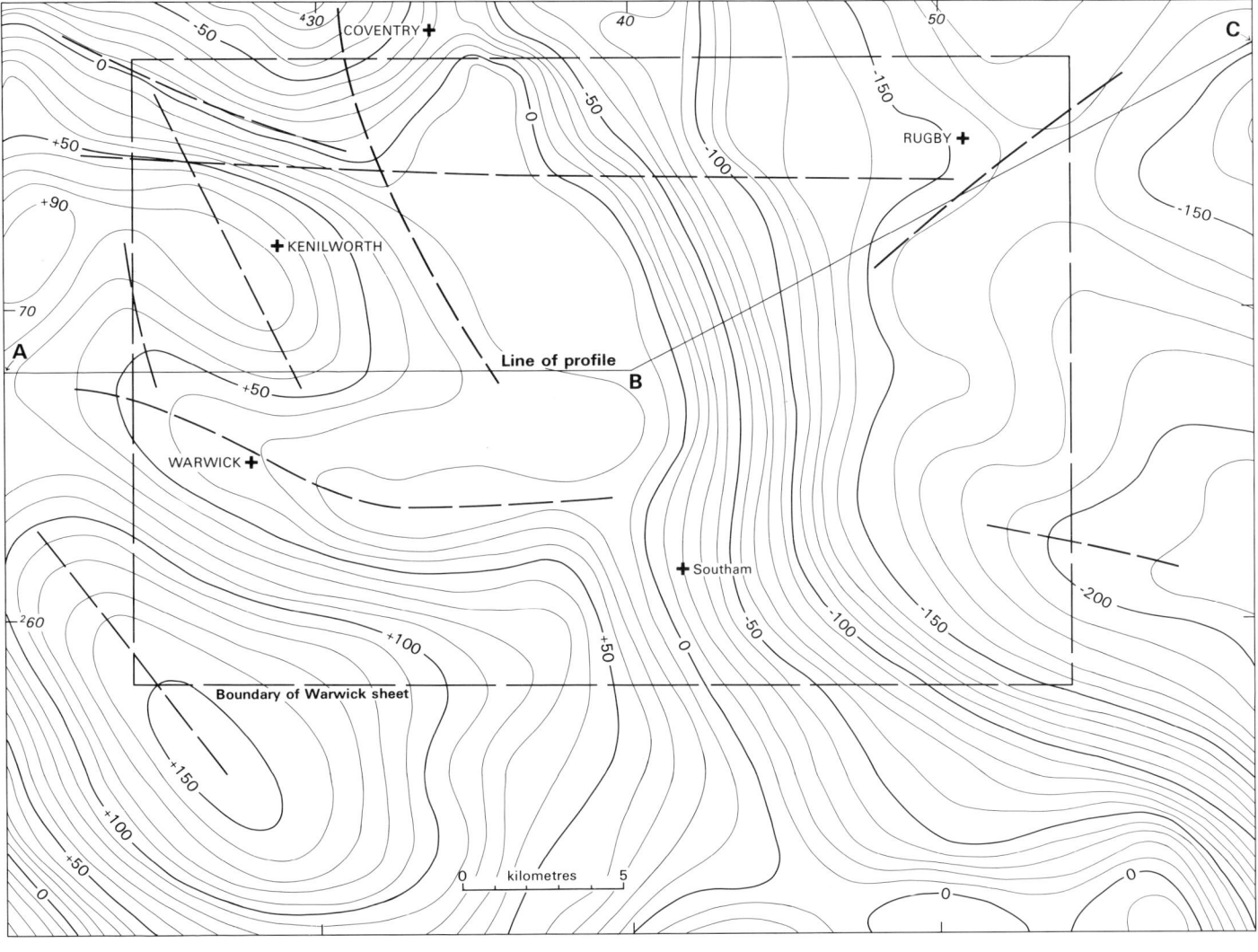

——— ——— Trend line

Figure 22 Aeromagnetic map

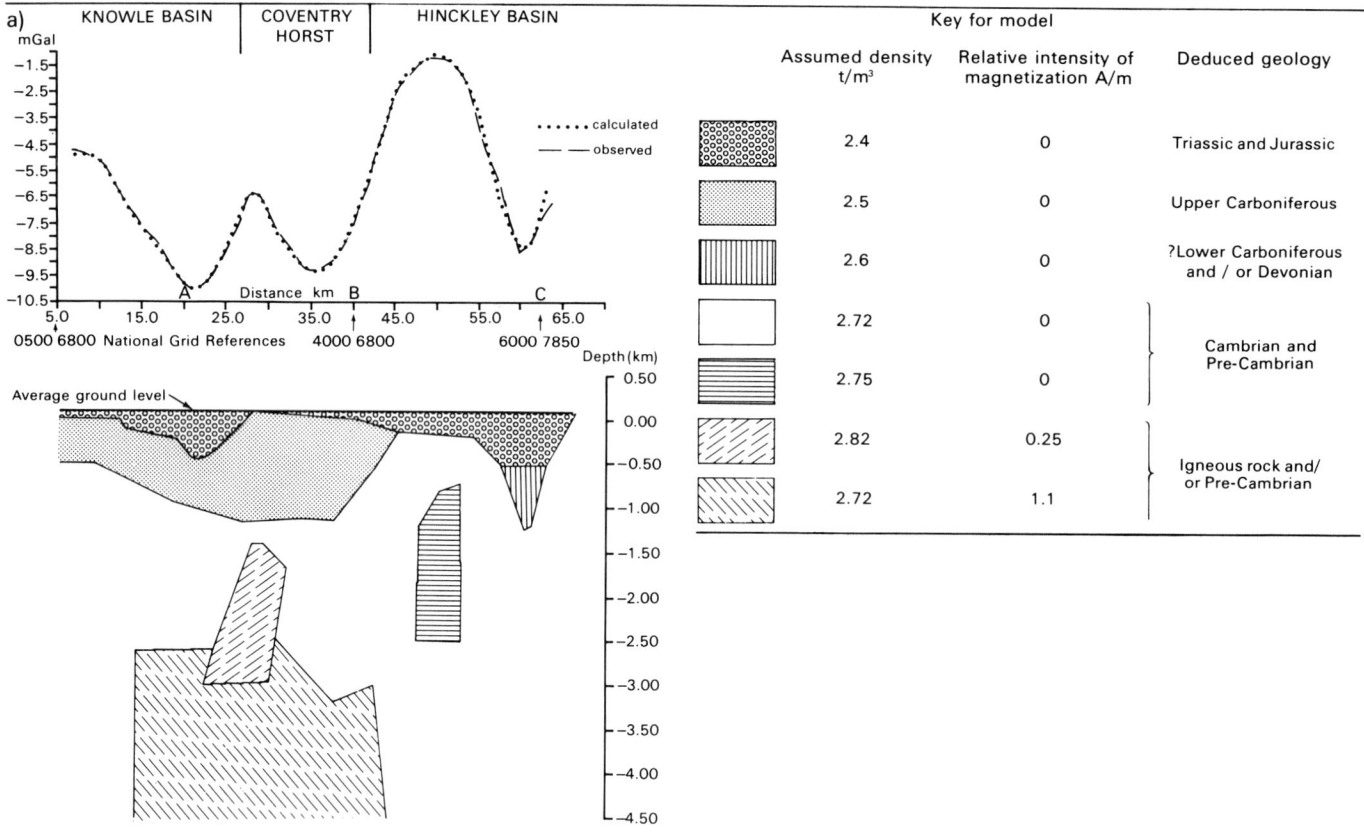

Figure 23a Gravity anomaly profile and model along line shown on Figure 20 with locating points A, B and C

NOTES

1 The regional gradient subtracted from the aeromagnetic profile values of Figure 22 is consistent with the background field expected for the British Isles (see Appendix 4). Bouguer anomaly values were taken directly as calculated for a reduction density of 2.4 t/m^3 to sea-level datum

2 The schematic model is a combination of density and magnetisation contrasts; it is reproduced at a smaller vertical scale below the aeromagnetic profile to emphasise the greater depth of the structure interpreted from the magnetic data

3 The model is controlled to the base of the Upper Carboniferous only within the Coventry Horst: the remainder is speculative and intended as a guide to the general structure. The thickness derived for the Trias basins flanking the horst depends upon the nature of the underlying deposits. Thus, if the Upper Carboniferous does not continue to the west beneath the Knowle Basin as shown, then the Trias is probably thicker. As the model reproduces the anomaly gradients, on a regional scale at least, it satisfies the criterion limiting the depth at which the main contrasts can occur.

4 Physical property values relevant to the Warwick district are discussed in Appendix 4. The background density of 2.72 t/m^3 was chosen to combine the Cambrian and Precambrian, though the density is expected to increase with depth from about 2.7 t/m^3 in the Cambrian. The block shown with a density of 2.75 t/m^3 below the Nuneaton Ridge is attributed to the Precambrian being at higher levels than elsewhere. The block of density 2.82 t/m^3 may indicate intrusive rock but its relation to the underlying, more strongly magnetised Precambrian is uncertain: it could be typical of the denser rocks which are thought to occur beneath the Warwickshire Coalfield north of Coventry.

Figure 23b Aeromagnetic profile and model along line shown on Figure 21 with locating points A, B and C

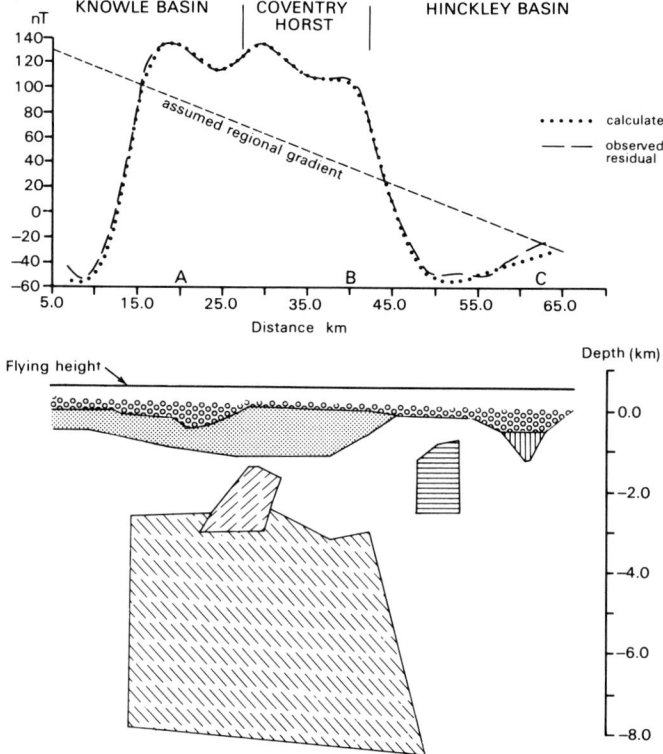

the south-west. An anomaly low attributed to the Knowle Basin is developed 15 km due west from that near Offchurch, with a north–south contour alignment near the margin of the district that may represent further thickening of the Trias into the basin across the extrapolation of the Western Boundary Fault. There is no geophysical evidence to suggest that the Upper Carboniferous sequence of the South Warwickshire Coalfield does not continue westwards beneath the Knowle Basin, although several quite different geological models are compatible with the available data. Seismic reflection records shown a zone of intense disturbance between the Warwick and Western Boundary faults across which is it not possible to trace the Carboniferous marker horizons. The block of high magnetic anomaly values extends for 10 km west of the district suggesting continuity in the character of the basement on which these structures are superimposed. Discussion of the geophysical anomalies beyond the southern margin of the district is given in Edmonds and others (1965) and Williams and Whittaker (1974).   RMC

CHAPTER 9

Economic geology

BRICK CLAY

Bricks were formerly made from the thick basal mudstone of the Ashow Formation at Whitemoor [294 718] and Cherry Orchard [295 721]. Production of engineering bricks continued at Cherry Orchard until 1977. Most reserves of the mudstone have been sterilised by urban development. Farther east, Liassic clays and Wolston Clay have been exploited. Clays below the Blue Lias were extracted in conjunction with the underlying Langport Member limestone; there are old pits west of Rugby [417 706; 410 717; 432 732; 443 755; 462 769; 464 774; 459 772] and some that still contain remains of brick-kilns [4432 7549; 4613 7655]. Clays above the Blue Lias were worked in and south-west of Rugby at Cawston [465 728], New Bilton [492 756; ?495 757; ?499 757], Toft [476 700] and at the Rugby, or Dunchurch Road Pit [502 744] which closed in about 1876 (Oldham, 1879). Similar clays at Hillmorton were dug from the Upper Hillmorton Pit [527 738] and Lower Hillmorton Pit [533 739], the latter opened in about 1867 (Oldham, 1879). Wilson (1875) noted a disused claypit [532 740], and Thompson (1898) recorded both the Upper and Lower Hillmorton pits ('Mr Satchell's brickyard' and 'the Central Brickyard'), which were evidently working at that time, and the newer Red-Bank Brickworks [543 735]. All the Hillmorton pits seem to have closed before 1915 (Nuttall, 1916, pp.63, 64). The whole of the Middle Lias and the upper 25 m of the Lower Lias were worked at Napton-on-the-Hill [456 613]. Wolston Clay was dug for the making of bricks and tiles just north of Brandon [410 768] and probably near Princethorpe and Rugby.

BUILDING STONE

Sandstone from the Enville Group has been much used locally as a good-quality, durable freestone. Castle Quarry [2775 7190] provided much of the stone from which Kenilworth Castle was built, and there were several quarries near Love Lane [287 729]. Stoneleigh church is constructed of stone from the quarry on Motslow Hill [3295 7233], and the almshouses in Stoneleigh are fine examples of the use of Enville sandstone. Bromsgrove Sandstone has been worked at a number of quarries in and around Warwick, notably at Coton End Quarry [289 655], and several other large quarries lie between Bubbenhall and Willenhall [359 723; 362 736; 353 763; 365 772]. The Langport Member was quarried between Frankton and King's Newnham [416 706; 422 721; 431 732; 446; 762; 462 769; 459 773] and mined from a shaft just north of the district [4654 7772]. Much of the limestone was burnt to produce lime for agriculture and mortar, but its use as a building stone was mentioned by Oldham (1879, p.46) and examples include Little Lawford Hall [4692 7722], King's Newnham church [4481 7722] and

Limestone Hall [4412 7540]. Blue Lias limestone has been used locally for building, and was mixed with Langport Member limestone at Southam church [4179 6174].

CEMENT

The district is important for cement manufacture, with two large plants operated by the Rugby Portland Cement Company at Rugby [490 757] and Long Itchington [421 640]. Smaller works were formerly active at several other sites. The cement industry is based on the Blue Lias. Its origins can be traced back to the early part of the nineteenth century. There are small pits dating from that time at Draycote [449 706] and Lawford Heath [469 749]. Quarries along the old route of the Oxford Canal at Newbold on Avon [480 777 to 486 767] were active in the 1820s, and by the 1850s works had opened at Long Itchington, Stockton [445 643] and Harbury [395 587]. Work at Rugby started in 1865 and by the 1890s quarrying had begun at Newbold on Avon [495 770]. Originally work at the quarries was seasonal, ceasing in winter, but since 1925 operations have continued throughout the year.

The Blue Lias in bulk contains too high a ratio of silica to calcium for Portland cement manufacture. Until the 1930s the limestone beds ('cementstones') were hand-sorted, and most of the clay discarded. The material was crushed, formed into bricks, and stacked in rectangular kilns with alternate layers of coke. After firing, the resultant clinker was crushed and sold as cement. The rectangular kilns were later replaced by bottle-kilns and, in this century, by rotary kilns.

In order to make more efficient use of the raw materials a supply of high-calcium rock is needed to blend with the surplus clay. At Harbury, limestone from the Langport Member was used at first; later chalk from the Totternhoe Quarry near Dunstable, Bedfordshire, was transported to the works by rail. At Rugby a chalk/water slurry is now pumped through 92 km of pipeline from Kensworth Quarry, Bedfordshire, and thence a further 19 km to Long Itchington. A mixture of slurry and clay is fired in a rotary kiln, and this process has allowed use of the clay discarded in earlier years.

Currently, the Rugby Works produces 500 000 to 600 000 tonnes of cement each year; products comprise Ordinary Portland Cement, Sulphate-Resisting Portland Cement and, at Long Itchington, Rapid Hardening and Masonry Cement.

SAND AND GRAVEL

Most of the larger deposits of sand and gravel have been worked at some time; the notable exception is the Dunsmore Gravel which is probably too clayey and too poorly sorted. The older workings were mostly small pits opened for local use. Crofts (1982) has compiled a report on sand and gravel resources in the north of the district. The materials are used in the construction industry for the production of mortar and concrete.

Large areas of Baginton Sand and Gravel have been worked out, as around Brandon Wood, Ryton-on-Dunsmore, Baginton and the Cherry Orchard Brickworks, Kenilworth, but several pits are still active [363 716; 379 728; 392 737; 382 747]. Normally the sand and underlying gravel are extracted separately to minimise the necessity for grading. Large accessible reserves of Baginton Sand and Gravel remain around Bubbenhall and Brandon. The Hillmorton Sand east of Rugby has been extensively worked, mainly during the nineteenth century; all the pits are now disused, and the extent of urban development makes it unlikely that any commercial prospect remains.

Gravels of the Fourth Terrace have been worked at Barford Wood [285 620], together with underlying Baginton Sand and Gravel near Ryton-on-Dunsmore [364 759]. The latter locality also contains old pits in the First and Second terraces [391 752; 382 750]. Reserves of accessible sand and gravel in terrace deposits are greatest alongside the River Avon south of Warwick; they extend to several hundred hectares and are generally around 3 m thick, reaching 5–6 m in places. Alluvial fan gravels have been dug from two small pits [415 738; 413 727] near Stretton-on-Dunsmore.

COAL

With the exception of Binley Colliery all the coal workings to date in the Warwickshire Coalfield lie to the north of the district. At Binley component seams of the Thick Coal (mainly the High Main and Two yard) were worked prior to the closure of the colliery in 1963 (Chapter 3). Abandonment plans of the workings are held by the National Coal Board, and analytical data for the Two Yard are published (National Coal Board, 1957).

Extensions of the proven Warwickshire Coalfield to the west and south of Coventry were postulated by Eastwood and others (1923, p.47) and Mitchell (1942, p.17), and since 1972 the National Coal Board has carried out a major programme of exploratory drilling and seismic surveys in the South Warwickshire Prospect (Figure 7; National Coal Board, 1985). Although coal seams have been proved in all the Westphalian subdivisions (Chapter 3) the only recoverable reserves, amounting to 400 million tonnes, occur in the Thick Coal. In the Prime, Partial and Splitting Thick Coal areas (Figure 7) the coal would provide excellent domestic, industrial and power station fuels. To the south-east, however, the poorer quality of the coal makes it suitable only for power station fuel. In 1986 the National Coal Board announced its intention to proceed with a new mine to exploit these reserves from shafts at Hawkhurst Farm [265 795], just to the north of the district.

GYPSUM

During construction work in the 1960s, two mine shafts [4084 7290; 4086 7293] were discovered at School Lane,

Stretton-on-Dunsmore. They were bricklined, 1.5 and 2.3 m in diameter, about 30 m deep, and had been abandoned probably in the nineteenth century. Known locally as the 'plaster pits', the shafts evidently mark a gypsum mine or prospect, and gypsum debris occurs nearby. The shafts penetrate the upper part of the Mercia Mudstone, close below the Blue Anchor Formation, corresponding with the most gypsiferous part of the sequence proved in the Home Farm Borehole [4317 7309] and approximating to the horizon of the Tutbury and Newark gypsums of Staffordshire, Derbyshire and Nottinghamshire.

LIME

Lime for agriculture and mortar was formerly produced from limestone of the Langport Member, use of the stone for building being of secondary importance.   RAO, MGS

HYDROGEOLOGY AND WATER SUPPLY

The district lies mostly in hydrometric area 54, administered by the Severn-Trent Water Authority. Small areas in the south-east are within hydrometric areas 32 and 39, administered by the Anglian and the Thames water authorities respectively. Mean annual rainfall (1916–1950) ranges from about 650 mm in the north-east to more than 700 mm in the north-west; mean annual evaporation is about 500 mm. Since much of the district is underlain by relatively impermeable strata, seasonal variation in river flow is great.

Most public supplies are drawn from surface sources, the most important being river intakes at Royal Leamington Spa [329 657] and Eathorpe [388 688] on the River Leam, at Brownsover [515 765] on the River Avon, and at Cosford Canal [507 771] on the River Swift. Together with the impounding reservoir of Draycote Water [460 700], these surface sources are licensed for a maximum yield of about 88 million m^3/a. Groundwater resources have been discussed by Richardson (1928), Woodland (1942), Butler (1946) and Lyon (1949), and some generalised information has been published by the Severn River Authority (1974). The total licensed abstraction of groundwater for public supply in the district amounts to approximately 7.5 million m^3/a. The major aquifers lie within the Enville Group and the Bromsgrove Sandstone. Smaller supplies have been obtained from thin sandstones within the Mercia Mudstone, Liassic limestones, and superficial sands and gravels.

In the north-west of the district, groundwater is abstracted from sandstones and conglomerates of the Enville Group and from the upper part of the Keele Formation. A borehole of 300 mm diameter at Kenilworth [2953 7283] yielded 220 m^3/d. Mine shafts are commonly pumped in order to dewater coal workings, and large volumes of water discharged. A shaft at Newdigate [3346 8694], north of the district, yielded 1200 m^3/d. Such shafts remain useful sources, even when mining has ceased. Groundwater from Carboniferous sandstones is usually of good quality; total hardness ranges from 200 to 500 mg/l, and chloride ion concentration is generally less than 30 mg/l. However, water from coal mines or from boreholes in the immediate vicinity of coal mines may be of poor quality, with a chloride ion concentration rising to many hundreds of milligrammes per litre.

The most important aquifer in the district is the Bromsgrove Sandstone. A borehole [2899 6863] at Leek Wootton of 250 mm diameter penetrated about 120 m of Bromsgrove Sandstone and Enville Group and yielded about 400 m³/d. Another [3302 6906] at Cubbington, of 380 mm diameter, passed through almost a full thickness of Bromsgrove Sandstone into Enville Group and yielded nearly 440 m³/d. Large-diameter shafts give larger yields; the Lillington Well [3323 6806] gave over 2000 m³/d. Near or within the outcrop, the quality of water in the Bromsgrove Sandstone is good; total hardness is generally less than 300 mg/l, and chloride ion concentration less than 30 mg/l. With an increasing thickness of cover, quality deteriorates; total hardness (largely non-carbonate hardness) of up to 5400 mg/l has been recorded, and chloride ion concentrations of more than 1500 mg/l.

The Mercia Mudstone is largely impermeable but contains thin, impersistent sandstone beds which constitute minor local aquifers. Natural recharge is, however, restricted, and although initial yields may be good they usually decline with time. The water is generally very hard, and total hardness can exceed 600 mg/l.

Limestones of the Langport Member yield small supplies. A shaft of 11 m depth at Southam [4103 6189] which yielded about 650 m³/d was probably exceptional. Total hardness is commonly high, exceeding 380 mg/l.

The thin argillaceous limestones of the Lower Lias may contain a little water, but the quality is usually poor and the water saline.

Sands and gravels in the drift yield variable, usually small, supplies of groundwater: a 5 m shaft [476 729] at Cawston gave nearly 14 m³/d. The water tends to be hard, and the aquifers are particularly vulnerable to pollution from the ground surface, or from rivers or streams with which they may be in hydraulic continuity.

Mineral springs at Royal Leamington Spa (Richardson, 1928, pp.121–133) issue from the Bromsgrove Sandstone, although the water probably derives from gypsiferous beds in the overlying Mercia Mudstone. An analysis of water from these springs (Lyon, 1949) is as follows:

	mg/l
Total dissolved solids (at 180°C)	14 640
Total hardness (as $CaCO_3$)	3 410
Carbonate hardness (as $CaCO_3$)	124
Non-carbonate hardness (as $CaCO_3$)	3 286
Chloride (as Cl)	6 830

At present the New Well [3193 6552] is the sole source of supply to the Royal Pump Room and Baths.   PKM

REFERENCES

ALLEN, J. R. L. 1965. Fining upwards cycles in alluvial successions. *Geol. J.*, Vol. 4, 229–246.

AMBROSE, K. 1978. Stockton Locks Borehole. *Rep. Inst. Geol. Sci.*, No. 77/10, 3.

— and BREWSTER, J. 1979. Barby Borehole. *Rep. Inst. Geol. Sci.*, No. 79/12, 3–4.

— — 1982. A re-interpretation of parts of the 400 ft bench of southeast Warwickshire. *Quaternary Newsl.*, No. 36, 21–24.

— and IVIMEY-COOK, H. C. 1982. The Barby (IGS) Borehole near Daventry, Northamptonshire. *Rep. Inst. Geol. Sci.*, No. 82/1, 36–40.

ANON. 1900. Description of the fossil *Ichthyosaurus platydon* found at Stockton, 1898. *Rep. Rugby School Nat. Hist. Soc.* for 1899, 50.

ARKELL, W. J. 1933 *The Jurassic system in Great Britain.* 681pp. (Oxford: Clarendon Press.)

— 1947. The geology of Oxford. 267pp. Oxford.

ARTHURTON, R. S. 1980. Rhythmic sedimentary sequences in the Triassic Keuper Marl (Mercia Mudstone Group) of Cheshire, northwest England. *Geol. J.*, Vol. 15, 43–58.

BAILEY, W. 1582. A Briefe Discours of certain Bathes or Medicinal Waters in the countie of Warwicke, neere unto a village called Newham Regis. (London.)

BISHOP, W. W. 1958. The Pleistocene geology and geomorphology of three gaps in the Midland Jurassic escarpment. *Philos. Trans. R. Soc. London*, Series B, Vol. 241, 255–306.

BREWSTER, J. 1978. Harbury Quarry Borehole. *Rep. Inst. Geol. Sci.*, No. 77/10, 3.

BRITISH GEOLOGICAL SURVEY. 1985. Map 2: Contours on the top of the Pre-Permian surface of the United Kingdom (South). (Keyworth: British Geological Survey.)

BRODIE, P. B. 1875. The Lower Lias at Eatington and Kineton and on the Rhaetics in that neighbourhood and their further extension in Leicestershire, Nottinghamshire, Lincolnshire, Yorkshire and Cumberland. *Proc. Warwickshire Nat. Archeol. Field Club*, 3–14.

— and KIRSHAW, J. W. 1872. Excursion to Warwickshire. *Proc. Geol. Assoc.*, Vol. 2, 284–287.

BUCKLAND, W. 1823. *Reliquiae diluvianae.* (London: J. Murray.)

BUCKMAN, S. S. 1918. Jurassic chronology: I—Lias. *Q. J. Geol. Soc. London*, Vol. 73, 257–327.

BULMAN, O. M. B. 1954. The graptolite fauna of the Dictyonema Shales of the Oslo region. *Norsk Geol. Tidsskr.*, Vol. 33, 1–40.

— and RUSHTON, A. W. A. 1973. Tremadoc faunas from boreholes in central England. *Bull. Geol. Surv. G.B.*, No. 43, 1–40.

BUTLER, A. J. 1946. Water supply from underground sources of the Birmingham-Gloucester district, part III. *Geol. Surv. Wartime Pam.*, No. 32, 38–56.

CAVE, R. 1977. Geology of the Malmesbury district. *Mem. Geol. Surv. G.B.* 343pp.

CLEMENTS, R. G. (editor). 1975. *Report on the geology of Long Itchington Quarry.* 30pp. (Leicester: Department of Geology, Leicester University.)

— (editor). 1977. *Report on the geology of Parkfield Road Quarry, Rugby.* 54pp. (Leicester: Department of Geology, Leicester University.)

CLEMINSHAW, E. 1868a. On the natural history of the Rugby Lias. *Rep. Rugby School Nat. Hist. Soc.* for 1867, 32–37.

— 1868b. A list of local Lias fossils. *Rep. Rugby School Nat. Hist. Soc.* for 1867, 55–57.

COOK, A. H., HOSPERS, J. and PARASNIS, D. S. 1952. The results of a gravity survey in the country between Clee Hills and Nuneaton. *Q. J. Geol. Soc. London*, Vol. 107, 287–302.

COOPE, G. R. 1962. A Pleistocene coleopterous fauna with arctic affinities from Fladbury, Worcestershire. *Q. J. Geol. Soc. London*, Vol. 118, 103–123.

— 1968. An insect fauna from mid-Weichselian deposits at Brandon, Warwickshire. *Philos. Trans. R. Soc. London*, Vol. B, 254, 425–456.

COPE, J. C. W., GETTY, T. A., HOWARTH, M. K., MORTON, N., and TORRENS, H. S. 1980. A correlation of the Jurassic rocks of the British Isles. Part One: Introduction and Lower Jurassic. *Spec. Rep. Geol. Soc. London*, No. 14. 73pp.

COPE, K. G. and JONES, A. R. L. 1970. The Warwickshire Thick Coal and its mining environment. *C. R. 6e. Congr. Int. Stratigr. Geol. Carbonif., Sheffield 1967*, 585–598.

CORNWELL, J. D. and ALLSOP, J. M. 1981. Geophysical surveys in the Atherstone district. *Open File Reports British Geological Survey. Applied Geophysics Series*, No. 38

COX, H. M. M. 1953. The fossil plants of the Permian beds of England. Unpublished Ph.D. thesis, University of Cambridge.

CROFTS, R. G. 1982. The sand and gravel resources of the country between Coventry and Rugby, Warwickshire: description of 1:25 000 sheets SP 47 and part of SP 37. *Miner. Assess. Rep. Inst. Geol. Sci.*, No. 125.

DAVIES, W. and CAVE, R. 1976. Folding and cleavage determined during sedimentation. *Sediment. Geol.*, Vol. 15, 89–133.

DEAN, W. T., DONOVAN, D. T. and HOWARTH, M. K. 1961. The Liassic ammonite zones and subzones of the North-West European Province. *Bull. Br. Mus. Nat. Hist. (Geol.)*, Vol. 4, 433–505.

DIX, E. 1935. Note on the flora of the highest 'Coal Measures' of Warwickshire. *Geol. Mag.*, Vol. 72, 555–557.

DONOVAN, D. T., HORTON, A. and IVIMEY-COOK, H. C. 1979. The transgression of the Lower Lias over the northern flank of the London Platform. *J. Geol. Soc. London*, Vol. 136, 165–173.

DOUGLAS, T. D. 1974. The Pleistocene beds exposed at Cadeby, Leicestershire. *Trans. Leicestershire Lit. Philos. Soc.*, Vol. 68, 57–63.

— 1980. The Quaternary deposits of western Leicestershire. *Philos. Trans. R. Soc. London*, Series B, Vol. 258, 259–286.

DRINKWATER, B. C. 1912. Geological section. *Rep. Rugby School Nat. Hist. Soc.*, for 1911, 63–64.

DUNHAM, K. C. and POOLE, E. G. 1974. The Oxfordshire Coalfield. *Q. J. Geol. Soc. London*, Vol. 130, 387–392.

DURY, G. H. 1951. A 400-foot bench in south-eastern Warwickshire. *Proc. Geol. Assoc.*, Vol. 62, 167–173.

— 1952. The alluvial fill of the valley of the Warwickshire Itchen near Bishops Itchington. *Proc. Coventry Nat. Hist. Sci. Soc.*, Vol. 2, 180–185.

EASTWOOD, T., GIBSON, W., CANTRILL, T. C. and WHITEHEAD, T. H. 1923. The geology of the country around Coventry, including an account of the Carboniferous rocks of the Warwickshire Coalfield. *Mem. Geol. Surv. G.B.* 149pp.

EDMONDS, E. A., POOLE, E. G. and WILSON, V. 1965. Geology of the country around Banbury and Edge Hill. *Mem. Geol. Surv. G.B.* 137pp.

ELLIOTT, R. E. 1961. The stratigraphy of the Keuper Series in southern Nottinghamshire. *Proc. Yorkshire Geol. Soc.*, Vol. 33, 197–234.

EVANS, A. M. and MAROOF, S. I. 1976. Basement controls on mineralization in the British Isles. *Min. Mag.*, Vol. 134, 401–411.

FLEET, W. F. 1925. The chief heavy minerals in the rocks of the English Midlands. *Geol. Mag.*, Vol. 62, 98–128.

— 1927. The heavy minerals of the Keele, Enville, 'Permian', and Lower Triassic rocks of the Midlands, and the correlation of these strata. *Proc. Geol. Assoc.*, Vol. 38, 1–48.

FOX-STRANGWAYS, C. 1897. Geology of the London extension of the Manchester, Sheffield and Lincolnshire Railway. Part One: Annesley to Rugby. *Geol. Mag.*, Vol. 4, 49–59.

GETTY, T. A. 1973. A revision of the generic classification of the family Echioceratidae (Cephalopoda, Ammonoidea) (Lower Jurassic). *Palaeontol. Contrib. Kansas Univ.*, Vol. 63, 32pp.

GREEN, G. W. and MELVILLE, R. V. 1956. The stratigraphy of the Stowell Park Borehole (1949–1951). *Bull. Geol. Surv. G.B.*, No. 11, 1–66.

HAINS, B. A. and HORTON, A. 1975. *British regional geology: Central England.* (London: HMSO for Institute of Geological Sciences.)

HALLAM, A. 1960. A sedimentary and faunal study of the Blue Lias of Dorset and Glamorgan. *Philos. Trans. R. Soc. London*, Series B, Vol. 243, 1–44.

— 1964. Origin of the limestone–shale rhythm in the Blue Lias of England: A composite theory. *J. Geol.*, Vol. 72, 157–169.

— 1967. An environmental study of the Upper Domerian and Lower Toarcian in Great Britain. *Philos. Trans. R. Soc. London*, Series B, Vol. 252, 393–445.

— 1975. *Jurassic environments.* 269pp. (Cambridge University Press.)

— and BRADSHAW, M. J. 1979. Bituminous shales and oolitic ironstones as indicators of transgressions and regressions. *J. Geol. Soc. London*, Vol. 136, 157–164.

HAUBOLD, H. and KATZUNG, G. 1975. Die position der Autun/Saxon-Grenze (Unteres Perm) in Europa und Nordamerika. *Schriftnr. Geol. Wiss Berlin*, Vol. 3, 87–138.

— and SARJEANT, W. A. S. 1973. Tetrapodenfährten aus den Keele und Enville Groups (Permokarbon: Stefan und Autun) von Shropshire und South Staffordshire, Grossbritannien. *Z. Geol. Wiss Berlin*, Vol. 1, 895–933.

HORTON, A. and POOLE, E. G. 1977. The lithostratigraphy of three geophysical marker horizons in the Lower Lias of Oxfordshire. *Bull. Geol. Surv. G.B.*, No. 62, 13–23.

HOWARTH, M. K. 1958. The ammonites of the Liassic Family Amaltheidae. *Palaeontogr. Soc. (Monogr.)*

— 1978. The stratigraphy and ammonite fauna of the Upper Lias of Northamptonshire. *Bull. Br. Mus. Nat. Hist. (Geol.)*, Vol. 29, 235–288.

— 1980. The Toarcian age of the upper part of the Marlstone Rock Bed of England. *Palaeontology*, Vol. 23, 637–656.

HOWELL, H. H. 1859. The geology of the Warwickshire Coalfield and the Permian rocks and Trias of the surrounding district. *Mem. Geol. Surv. G.B.*

HULL, E. 1869. The Triassic and Permian rocks of the Midland Counties of England. *Mem. Geol. Surv. G.B.*

JACK, A. 1922. A palaeolithic implement from Barford, Warwickshire. *Proc. Prehist. Soc. E. Anglia*, Vol. 3, 621.

KENT, P. E. 1953. The Rhaetic beds of the north-east Midlands. *Proc. Yorkshire Geol. Soc.*, Vol. 29, 117–139.

— 1968. The Rhaetic Beds. 174–187 in *The geology of the East Midlands.* SYLVESTER-BRADLEY, P. C. and FORD, T. D. (editors). (Leicester: Leicester University Press.)

KELLY, M. R. 1968. Floras of Middle and Upper Pleistocene age, from Brandon, Warwickshire. *Philos. Trans. R. Soc. London*, Series B, Vol. 254, 401–415.

KEULEGAN, G. H. and KRUMBEIN, W. C. 1949. Stable configuration of bottom slope in a shallow sea, and its bearings on geological processes. *Trans. Am. Geophysics Union*, Vol. 30, 855–861.

LAWE, J. D. 1869. Geological notices. *Rep. Rugby School Nat. Hist. Soc.* for 1868, 43, 44.

LE BAS, M. J. 1972. Caledonian igneous rocks beneath central and eastern England. *Proc. Yorkshire Geol. Soc.*, Vol. 39, 71–86.

LOWE, W. B. 1873. Explanation of section of Victoria Lime Works. *Rep. Rugby School Nat. Hist. Soc.* for 1872, 48, 49.

LYON, A. L. 1949. The hydrogeology of the Coventry District. *J. Inst. Water. Eng.*, Vol. 3, No. 3, 209–260.

MAIDWELL, F. T. 1910. On the basement bed of the Keuper. *Proc. Warwickshire Nat. Arch. Field Club* (for 1910), 5–10.

MASSON SMITH, D., HOWELL, P. M., ABERNETHY CLARK, A. B. D. E. and PROCTOR, D. W. 1974. The National Gravity Reference Net, 1973. *Prof. Pap. Ordnance Surv. G.B.*, No. 20.

MITCHELL, G. H. 1942. The geology of the Warwickshire Coalfield. *Geol. Surv. Wartime Pam.*, No. 25. 42pp.

MURCHISON, R. I. and STRICKLAND, H. E. 1840. On the upper formation of the New Red Sandstone System in Gloucestershire, Worcestershire and Warwickshire. *Trans. Geol. Soc. London*, Vol. 5 (2nd Ser.), 331–348.

MYKURA, H. and HAMPTON, B. P. 1984. On the mechanism of formation of reduction spots in the Carboniferous/Permian red beds of Warwickshire. *Geol. Mag.*, Vol. 121, 71–74.

NATIONAL COAL BOARD. 1957. Warwickshire Coalfield seam maps. *NCB Sci. Dep. Coal Surv.*

— 1985. The South Warwickshire Prospect: A consultation paper. (NCB: South Midlands Area.)

NUTTALL, W. L. F. 1916. Geological section. *Rep. Rugby School Nat. Hist. Soc.* for 1915, 56–85.

OLD, R. A. 1982. Geological notes and local details for 1:10 000 sheets: SP 17 NE (Solihull and Knowle). (Keyworth: Institute of Geological Sciences.)

OLDHAM, T. B. 1878. List of Rugby fossils. *Rep. Rugby School Nat. Hist. Soc.* for 1877, 50–54.

— 1879. Geology of the neighbourhood of Rugby. *Rep. Rugby School Nat. Hist. Soc.* for 1878, 39–46.

— and JONES, E. 1879. Notes on the White Lias in the neighbourhood of Rugby. *Rep. Rugby School Nat. Hist.* for 1877, 48–54.

ORBELL, G. 1973. Palynology of the British Rhaeto-Liassic. *Bull. Geol. Surv. G.B.*, No. 44, 1–44.

OSBORNE, P. J. and SHOTTON, F. W. 1968. The fauna of the channel deposit of early Saalian age, at Brandon, Warwickshire. *Philos. Trans. R. Soc. London*, Vol. B 254, 417–424.

OWENS, R. M., FORTEY, R. A., COPE, J. C. W., RUSHTON, A. W. A. and BASSETT, M. G. 1982. Tremadoc faunas from the Carmarthen district, South Wales. *Geol. Mag.*, Vol. 119, 1–38.

PARASNIS, D. S. 1952. A study of rock densities in the English Midlands. *Mon. Nottingham Astr. Soc. Geophys. Suppl.*, No. 6, 252–271.

PATON, R. L. 1974. Lower Permian Pelycosaurs from the English Midlands. *Palaeontology*, Vol. 17, 541–552.

— 1975. A Lower Permian Temnospondylus amphibian from the English Midlands. *Palaeontology*, Vol. 18, 831–845.

PLAYLE, B. 1962. The Heathcote Mammoth tusk. *Warwickshire Nat. Hist. Soc.*, 8th Annual Report, 20.

POOLE, E. G. 1969. The stratigraphy of the Geological Survey Apley Barn borehole, Witney, Oxfordshire. *Bull. Geol. Surv. G.B.*, No. 29, pp.103.

— 1977. The stratigraphy of the Steeple Aston Borehole, Oxfordshire. *Bull. Geol. Surv. G.B.* No. 57, 85pp.

— 1978. The stratigraphy of the Withycombe Farm Borehole. *Bull. Geol. Surv. G.B.*, No. 68, 63pp.

— WILLIAMS, B. J. and HAINS, B. A. 1968. Geology of the country around Market Harborough. *Mem. Geol. Surv. G.B.* 92pp.

RAMSBOTTOM, W. H. C., CALVER, M. A., EAGER, R. M. C., HUDSON, F., HOLLIDAY, D. W., STUBBLEFIELD, C. J. and WILSON, R. B. 1978. A correlation of Silesian rocks in the British Isles. *Spec. Rep. Geol. Soc. London*, No. 10. 82pp.

RICE, R. J. 1968. The Quaternary deposits of central Leicestershire. *Philos. Trans. R. Soc. London*, Series B, Vol. 262, 459–509.

— 1981. The Pleistocene deposits of the area around Croft in South Leicestershire. *Philos. Trans. R. Soc. London*, Series B, Vol. 293, 385–417.

RICHARDSON, L. 1912. The Rhaetic rocks of Warwickshire. *Geol. Mag.*, Vol. 9, 24–53.

— 1928. The wells and springs of Warwickshire. *Mem. Geol. Surv. G.B.* 204pp.

— and FLEET, W. F. 1926. On sandstones with breccias below the Trias at Stratford-on-Avon and elsewhere in South Warwickshire. *Proc. Geol. Assoc.*, Vol. 37, 283–305.

RUSHTON, A. W. A. 1981. A polymorphic graptolite from concealed Tremadoc rocks of England. *Geol. Mag.*, Vol. 118, 615–622.

SEVERN RIVER AUTHORITY. 1974. First periodical survey of water resources and demands. Severn River Authority, Malvern, Worcs. 241pp.

SHOTTON, F. W. 1927. The conglomerates of the Enville Series of the Warwickshire Coalfield. *Q. J. Geol. Soc. London*, Vol. 83, 604–621.

— 1929. The geology of the country around Kenilworth (Warwickshire). *Q. J. Geol. Soc. London*, Vol. 85, 167–222.

— 1932. An exposure of contorted drift on the London Road. *Proc. Coventry Nat. Hist. Soc.*, Vol. 1, 52–53.

— 1933. New evidence on the origin of breccias and conglomerates in the Warwickshire Coalfields: The Mount Nod Boreholes, Coventry. *Geol. Mag.*, Vol. 70, 466–476.

— 1953. The Pleistocene deposits of the area between Coventry, Rugby and Leamington, and their bearing on the topographic development of the Midlands. *Philos. Trans. R. Soc. London*, Series B, Vol. 237, 209–260.

— 1968. The Pleistocene succession around Brandon, Warwickshire. *Philos. Trans. R. Soc. London*, Series B, Vol. 254, 387–400.

— 1976a. Amplification of the Wolstonian stage of the British Pleistocene. *Geol. Mag.*, Vol. 113, 241–250.

— 1976b. Lion in the No. 2 terrace of the River Avon. *Proc. Coventry Nat. Hist. Sci. Soc.*, Vol. 4, 323–324.

— 1977. The English Midlands. Guidebook for excursion A2, Tenth INQUA Congress Norwich, Geo Abstracts Ltd. 51pp.

— 1983. The Wolstonian Stage of the British Pleistocene in and around its type area of the English Midlands. *Quaternary Sci. Rev.*, Vol. 2, 261–280.

— BANHAM, P. H. and BISHOP, W. W. 1977. Glacial–interglacial stratigraphy of the Quaternary in Midland and Eastern England. In *British Quaternary Studies*. SHOTTON, F. W. (editor). (Oxford: Clarendon Press.)

— and WEST, R. G. 1969. Stratigraphic table of the British Quaternary. 155–157 *in* Recommendations on a stratigraphic usage. *Proc. Geol. Soc. London*, No. 1656.

SMITH, D. B., BRUNSTROM, R. G. W., MANNING, P. I., SIMPSON, S. and SHOTTON, F. W. 1974. A correlation of Permian Rocks in the British Isles. *Spec. Rep. Geol. Soc. London*, No. 5. 45pp.

SPENS, W. 1899. Geological section. *Rep. Rugby School Nat. Hist. Soc.* for 1898, 82–84.

STUBBLEFIELD, C. J. and BULMAN, O. M. B. 1927. The Shineton Shales of the Wrekin district. *Q. J. Geol. Soc. London*, Vol. 83, 96–146.

SUMBLER, M. G. 1980. Home Farm Borehole, Stretton-on-Dunsmore, 4–6 in *Rep. Inst. Geol. Sci.*, No. 79/12.

— 1981. Coventry (169) and Warwick (184) sheets. In *Rep. Inst. Geol. Sci.*, No. 80/11.

— 1983a. A new look at the type Wolstonian glacial deposits of Central England. *Proc. Geol. Assoc.*, Vol. 94, 23–31.

— 1983b. The type Wolstonian sequence — some further comments. *Quaternary Res. Assoc. Newsl.*, No. 40, 36–39.

SYLVESTER-BRADLEY, P. C. and FORD, T. D. (editors). 1968. *The geology of the East Midlands.* 400pp. (Leicester: Leicester University Press.)

TAYLOR, K. and RUSHTON, A. W. A. 1971. The pre-Westphalian geology of the Warwickshire Coalfield. *Bull. Geol. Surv. G.B.*, No. 35. 150pp.

THOMPSON, B. 1880. Notes on local geology, Part IV. The Lias System. *J. Northamptonshire Nat. Hist. Soc.*, Vol. 1, 142–149.

— 1889. *The Middle Lias of Northamptonshire.* 149pp. (London: Simpkin, Mashall & Co.)

— 1898. Excursion to Hill Morton and Rugby. *Proc. Geol. Assoc.*, Vol. 15, 428–433.

— 1899. Geology of the Great Central Railway (New extension to London of the Manchester, Sheffield and Lincolnshire Railway): Rugby to Catesby. *Q. J. Geol. Soc. London*, Vol. 55, 65–88.

TOMES, R. F. 1878. On the stratigraphical position of corals of the Lias of the Midland and Western counties of England and South Wales. *Q. J. Geol. Soc. London*, Vol. 34, 179–195.

TOMLINSON, M. E. 1925. The river terraces of the lower valley of the Warwickshire Avon. *Q. J. Geol. Soc. London*, Vol. 81, 137–169.

— 1935. The superficial deposits of the country north of Stratford on Avon. *Q. J. Geol. Soc. London*, Vol. 91, 423–460.

WALKER, A. D. 1969. The reptile fauna of the 'Lower Keuper' Sandstone. *Geol. Mag.*, Vol. 106, 470–476.

WARRINGTON, G. 1970. The stratigraphy and palaeontology of the 'Keuper' series of the Central Midlands of England. *Q. J. Geol. Soc. London*, Vol. 126, 183–223.

— AUDLEY-CHARLES, M. G., ELLIOTT, R. E., EVANS, W. B., IVIMEY-COOK, H. C., KENT, P., ROBINSON, P. L., SHOTTON,

F. W. and TAYLOR, F. M. 1980. A correlation of Triassic rocks in the British Isles. *Spec. Rep. Geol. Soc. London*, No. 13: 78pp.

WHITEHEAD, T. H., ANDERSON, W., WILSON, V. and WRAY, D. A. 1952. The Liassic Ironstones. *Mem. Geol. Surv. G.B.* 211pp.

WHITTAKER, A. and GREEN, G. W. 1983. Geology of the country around Weston-super-Mare. *Mem. Geol. Surv. G.B.* 147pp.

WILLIAMS, B. J. and WHITTAKER, A. 1974. Geology of the country around Stratford-upon-Avon. *Mem. Geol. Surv. G.B.* 127pp.

WILLS, L. J. 1938. The Pleistocene development of the Severn from Bridgenorth to the sea. *Q. J. Geol. Soc. London*, Vol. 94, 161–242.

— 1970. The Triassic succession in the central Midlands in its regional setting. *Q. J. Geol. Soc. London*, Vol. 126, 225–285.

WILSON, J. M. 1869. Rugby Waterworks. Remarks to accompany the section of the well. *Rep. Rugby School Nat. Hist. Soc.* for 1868, 41–42.

— 1870a. On the drifts and gravels and alluvial soils of Rugby and its neighbourhood. *Rep. Rugby School Nat. Hist. Soc.* for 1869, 16–35.

— 1870b. On the surface deposits in the neighbourhood of Rugby. *Q. J. Geol. Soc. London*, Vol. 26, 192–202.

— 1874. The Rugby Drift. *Rep. Rugby School Nat. Hist. Soc.* for 1873, 10–13.

— 1875. Contributions to the geology of Hillmorton. *Rep. Rugby School Nat. Hist. Soc.* for 1874, 8–13.

— 1876. Note on some bones found in a drift at Lawford. *Rep. Rugby School Nat. Hist. Soc.* for 1875, 73, 74.

WOODLAND, A. W. 1942. Water supply from underground sources of the Oxford-Northampton district, part III. *Geol. Surv. Wartime Pam.*, No. 4, 4–7.

WOODWARD, H. B. 1893. The Jurassic rocks of Britain, Vol. III. The Lias of England and Wales (Yorkshire excepted). *Mem. Geol. Surv. G.B.* 399pp.

— 1897. Geology of the London extension of the Manchester, Sheffield and Lincolnshire Railway, Part II: Rugby to Quainton Road, near Aylesbury. *Geol. Mag.*, Vol. 34, 97–105.

WORSSAM, B. C. and OLD, R. A. *In press.* Geology of the country around Coalville. *Mem. Geol. Surv. G.B.*.

WYATT, R. J. 1982. Geological notes and local details for 1:10 000 sheets: SO 93 NW and SW (Bredon and Ashchurch). (Keyworth: Institute of Geological Sciences.)

APPENDIX 1

Borehole catalogue

The most important boreholes in the district are listed below under the appropriate National Grid 1:10 000 quarter sheets, and references to full accounts are also given. The borehole numbers are those of the BGS 6-inch records system. Boreholes marked * are held 'Commercial in Confidence'.

SP 26 NW

63   County Mental Hospital, Hatton. Mercia Mudstone, Bromsgrove Sandstone [2498 6715]. Richardson 1928, pp.82–83.

SP 26 NE

21   Woodloes Well. Enville Group [2768 6642]. Richardson 1928, p.147.

23   Austin Edwards Ltd. Enville Group [2867 6539]. Richardson 1928, p.149.

24   Emscote Mills. Bromsgrove Sandstone, Enville Group [2906 6556]. Richardson 1928, p.148.

25a   Warwick Laundry. Bromsgrove Sandstone, Enville Group [2919 6526].

26b   Warwick Dyeworks. Bromsgrove Sandstone, Enville Group [2915 6526].

*87a   Woodcote Lane (NCB). Triassic, Permian, Westphalian, Cambrian [2818 6947].

*88   Wedgnock (NCB). Triassic, Permian, Westphalian, Cambrian [2705 6726].

SP 26 SE

79   Budbrooke Barracks. Mercia Mudstone, Bromsgrove Sandstone, Enville Group [2546 6471]. Butler 1946, p.49.

*95   Barford (NCB). Triassic, Permian, Westphalian, Cambrian [2384 6209].

SP 27 NE

1   Tile Hill Railway Station. Enville Group [2781 7755]. Butler 1946, pp.38–39.

*7   Ten Shilling Wood (NCB). Westphalian, Cambrian [2934 7683].

*8   Beanit Spinney (NCB). Westphalian, Cambrian [2655 7658].

*48   Westwood Heath (NCB). Westphalian, Cambrian [2810 7659].

*49   Catchems Corner (NCB). Westphalian, Cambrian [2533 7634].

*50   Bockenden (NCB). Westphalian, Cambrian [2802 7525].

*52   Black Waste Wood (NCB). Permian, Westphalian, Cambrian [2745 7595].

*53   Hurst Farm (NCB). Westphalian, Cambrian [2890 7587].

SP 27 SE

13   Kenilworth Waterworks (4 wells). Enville Group [2953 7283]. Richardson 1928, pp.90–94.

14   Holly Cottage, Kenilworth. Enville Group [2830 7323].

*16   Rouncil Lane (NCB). Triassic, Permian, Westphalian, Cambrian [2643 7024].

*17   Little Chase (NCB). Triassic, Permian, Westphalian, Cambrian [2646 7305].

*18   Redfern Farm (NCB). Triassic, Permian, Westphalian, Cambrian [2526 7479].

*19   Crackley Wood (NCB). Permian, Westphalian, Cambrian [2912 7480].

*28   Southurst Farm (NCB). Permian, Westphalian, Cambrian [2874 7378].

*29   Fernhill (NCB). Triassic, Permian, Westphalian, Cambrian [2510 7047].

*30   Long Meadow Wood (NCB). Westphalian, Cambrian [2716 7415].

31   Harbury Quarry (BGS). Lower Lias, Penarth Group, Mercia Mudstone [3992 5899]. IGS 1978, p.3.

SP 36 NW

7   LMS Engine Shed, Royal Leamington Spa. Mercia Mudstone, Bromsgrove Sandstone, Enville Group [3043 6606]. Butler 1946, pp.46–47.

14   Sewage Works, Royal Leamington Spa. Mercia Mudstone, Bromsgrove Sandstone, Enville Group [3092 6539]. Richardson 1928, pp.118–119.

16   Campion Terrace, Royal Leamington Spa. Mercia Mudstone, Bromsgrove Sandstone, Enville Group [3245 6632]. Richardson 1928, pp.114–116.

17   Leicester Lane, Cubbington. Bromsgrove Sandstone, Enville Group [3302 6906].

18   Lillington Well. Glacial drift, Bromsgrove Sandstone [3323 6806]. Richardson 1928, pp.116–118.

27b   Leek Wooton. Bromsgrove Sandstone, Enville Group [2899 6863].

*32   Cubbington Heath (NCB). Triassic, Westphalian, Cambrian [3380 6976].

33   Cubbington (BGS). Dunsmore Gravel, Upper Wolston Clay, Wolston Sand, Lower Wolston Clay, Thrussington Till [3493 6917].

*82   Sandy Lane (NCB). Triassic, Permian, Westphalian, Cambrian [3058 6757].

*86   Weston (NCB). Triassic, ?Permian, Westphalian, Cambrian [3483 6890].

SP 36 NE

5   Weston Colony. Mercia Mudstone, Bromsgrove Sandstone, Enville Group [3663 6907]. Butler 1946, p.44.

*8   Chalet (NCB). Triassic, Westphalian, Cambrian [3694 6698].

9   Offchurch No. 1 (BGS). Dunsmore Gravel, Upper Wolston Clay, Wolston Sand and Gravel, Thrussington Till, Mercia Mudstone [3680 6545].

11   Offchurch No. 2a (BGS). Oadby Till, Upper Wolston Clay, Wolston Sand and Gravel [3623 6538].

12   Offchurch No. 3 (BGS). Upper Wolston Clay, Wolston Sand and Gravel, Thrussington Till [3582 6524].

13   Hunningham (BGS); Dunsmore Gravel, Upper Wolston Clay, Wolston Sand and Gravel, Lower Wolston Clay, Thrussington Till [3761 6724].

14   Weston under Wetherley (BGS). Thrussington Till, Mercia Mudstone [3618 6911].

15   Fosse Farm (BGS). Dunsmore Gravel, Upper Wolston Clay, Wolston Sand and Gravel [3768 6638].

*16   Eathorpe (NCB). Triassic, Westphalian, Cambrian [3912 6776].

SP 36 SW

13   Tachbrook Mallory. Mercia Mudstone, Bromsgrove Sandstone, ?Enville Group [3226 6225]. Richardson 1928, pp.81–82.

14   Heathcote. Mercia Mudstone, Bromsgrove Sandstone, Enville Group [3094 6330]. Richardson 1928, p.149.

16   Radford Semele. Drift, Mercia Mudstone,
      Bromsgrove Sandstone [3459 6408]. Butler 1946,
      p.52.

17   Royal Leamington Spa Steam Laundry. Mercia
      Mudstone, Bromsgrove Sandstone [3212 6444].
      Richardson 1928, pp. 119–120.

SP 36 SE

8   Harbury Hall. Lower Lias, Penarth Group, Mercia
      Mudstone, [3760 4007]. Richardson 1928, p.60.

10   Southam Road Well. Oadby Till, Wolston Sand
      [3505 6430]. Richardson 1928, p.85.

11   Radford Semele No. 1 (BGS). Dunsmore Gravel,
      ?Upper Wolston Clay, ?Wolston Sand and Gravel,
      ?Lower Wolston Clay, Mercia Mudstone [3560
      6382].

12   Radford Semele No. 2 (BGS). ?Lower Wolston
      Clay, Thrussington Till [3551 6395].

13   Radford Semele No. 3 (BGS). Head, Thrussington
      Till, Mercia Mudstone [3517 6390].

14   Radford Semele No. 4 (BGS). Head, ?Lower
      Wolston Clay, Thrussington Till, Mercia Mudstone
      [3518 6369].

˙19   Middle Road (NCB). Triassic, Westphalian,
      Cambrian [3512 6155].

˙20   Ufton (NCB). Triassic, Westphalian, Cambrian
      [3844 6425].

SP 37 NW

136   Green Lane. Enville Group [3220 7598]. Butler
      1946, pp.39–40.

˙312 Cryfield Grange (NCB). Westphalian, Cambrian
      [3031 7532].

SP 37 NE

17   Binley Colliery No. 17. Triassic, Westphalian [3829
      7730].

18   Binley Colliery No. 18. Triassic, Westphalian [3803
      7733].

21   Willenhall Bridge. Bromsgrove Sandstone, Enville
      Group [3584 7670]. Richardson 1928, pp.182–183.

22   Whitley Well. Bromsgrove Sandstone, Enville Group
      [3577 7673]. Richardson 1928, pp.183.

23   Ryton No. 2 (NCB). Triassic, Westphalian, Cam-
      brian [3947 7562].

24   Ryton No. 3 (NCB). Triassic, Westphalian, dolerite
      [3694 7531].

25   Humber Co. Folly Lane. Enville Group [3513 7825].
      Richardson 1928, pp.190–191.

26   Brandon No. 10 (NCB). Triassic, Westphalian [3873
      7641].

27   Brandon No. 11 (NCB). Triassic, Westphalian [3907
      7543].

104   Binley Colliery No. 2 Shaft. Triassic, Westphalian
      [3795 7728].

105   Beddows Mine, Binley Colliery. Middle Coal
      Measures [3728 7715].

138, 139 Whitley (BGS). Baginton Sand and Gravel, Broms-
          grove Sandstone (Bubbenhall Clay absent) [3542
          7701; 3536 7707].

˙398 Rowley Road (NCB). Triassic, Westphalian, Cam-
      brian [3506 7510].

399–402 BGS Industrial Minerals Assessment Unit. Crofts
         1982.

SP 37 SW

˙60   Wainbody Wood (NCB). Permian, Westphalian,
      Cambrian [3139 7419].

˙100 Ashow (NCB). Permian, Westphalian, Cambrian
      [3053 7161].

˙101 Black Spinney (NCB). Westphalian, Cambrian [3436
      7326].

˙109 Millburn Grange (NCB). Permian, Westphalian,
      Cambrian [3041 7379].

˙110 Crewe Farm (NCB). Permian, Westphalian, Cam-
      brian [3136 7200].

˙111 Stareton (NCB). Permian, Westphalian, Cambrian
      [3320 7218].

112, 113 BGS Industrial Minerals Assessment Unit. Crofts
          1982.

SP 37 SE

1   Ryton No. 12 (NCB). Triassic, Westphalian [3922
      7385].

2, 3 Ryton Nos. 5 and 5a (NCB). Triassic, Westphalian
      [3916 7431; 3915 7430].

28   Ryton No. 8 (NCB). Triassic, Westphalian [3923
      7466].

29   Baginton. Triassic, Westphalian (Enville Group)
      [3640 7410]. Butler 1946, pp.41–42.

30   Ryton No. 4 (NCB). Triassic, Westphalian [3872
      7491].

32   Ryton No. 1 (NCB). Triassic, Westphalian, Cam-
      brian [3731 7397].

33   Ryton No. 6 (NCB). Triassic, Westphalian [3889
      7362].

34   Ryton No. 7 (NCB). Triassic, Westphalian [3906
      7357].

˙35   Rock Farm (NCB). Triassic, Westphalian, Cam-
      brian [3644 7628].

177, 178 Bubbenhall (BGS). Baginton Sand and Gravel, Mer-
          cia Mudstone (Bubbenhall Clay absent) [3583 7512;
          3578 7168]

308   Burnthurst Farm. Wolstonian [3880 7158]. IGS
      1980, p.2.

˙309 Weston Fields (NCB). Triassic, Westphalian, Cam-
      brian [3632 7153].

˙310 Burnhurst (NCB). Triassic, Westphalian, Cambrian
      [3919 7170].

311–323 BGS Industrial Mineral Assessment Unit. Crofts
         1982.

SP 45 NW

1   Napton No. 7 (Gas Council). Lower Lias, Penarth
      Group (Gamma logs only) [4427 5931].

˙6   Ladbroke (NCB). Jurassic, Triassic, Westphalian,
      Cambrian [4164 5958].

SP 45 NE

1   Napton No. 9 (Gas Council). Lower Lias (Gamma
      logs only) [4629 5833].

SP 46 NW

2   Birdingbury Hall. Blue Lias [4322 6869]. Richard-
      son 1928, p.49.

˙4   Fieldhouse Farm (NCB). Jurassic, Triassic,
      Westphalian, Cambrian [4383 6635].

˙5   Princethorpe Meadows (NCB). Triassic,
      Westphalian, Cambrian [4046 6997].

˙6   Newfields Farm (NCB). Triassic, Westphalian,
      Cambrian [4041 6689].

˙7   Birdingbury (NCB). Triassic, Westphalian, Cam-
      brian [4195 6768].

SP 46 SW

1   Napton No. 1 (Gas Council). Lower Lias and
      Penarth Group (Gamma log only) [4494 6116].

7   Southam Waterworks. Lower Lias and Penarth
      Group [4103 6189]. Richardson 1928, pp.62–63.

8   Long Itchington Cement Works. Jurassic, Triassic
      and ?Westphalian [4199 6396].

9   Stockton Locks (BGS). Lower Lias, Penarth Group,
      Mercia Mudstone [4297 6485]. IGS 1977, p.3.

˙14   Southam (NCB). Jurassic, Triassic, Westphalian,
      Cambrian [4200 6334].

*15   Gibraltar (NCB). Jurassic, Triassic, Westphalian, Cambrian [4461 6465].

SP 46 SE
1–5   Napton Nos. 2–5, 12, (Gas Council). Napton on the Hill. Lower Lias and Penarth Group (Gamma logs only).
*8   Napton Fields (NCB). Jurassic, Triassic, Westphalian, Cambrian [4514 6154].
*9   In Meadow Gate (NCB). Jurassic, Triassic, Westphalian, Cambrian [4850 6036].

SP 47 NW
1   Priory Hill, Wolston. Drift, Mercia Mudstone, Bromsgrove Sandstone [4234 7593]. Richardson 1928, pp.54–55.
63–74   BGS Industrial Minerals Assessment Unit. Crofts 1982.

SP 47 NE
64–69   BGS Industrial Minerals Assessment Unit. Crofts 1982.

SP 47 SW
72   Home Farm (BGS). Lower Lias, Penarth Group, Mercia Mudstone, Bromsgrove Sandstone, Tremadoc [4317 7309]. IGS 1979, pp.4–6.
73–87   BGS Industrial Minerals Assessment Unit. Crofts 1982.

SP 47 SE
24–30   BGS Industrial Minerals Assessment Unit. Crofts 1982.

SP 57 NW
92   British Thomson-Houston Co. Ltd. Lower Lias, Penarth Group, Mercia Mudstone [5100 7646]. Richardson 1928, pp.96–97.
202   Vicarage Hill (BGS), Clifton upon Dunsmore. Dunsmore Gravel, Oadby Till [5244 7598]. IGS 1980, p.3.

SP 57 SW
3   Rugby Waterworks. Lower Lias, Penarth Group, Mercia Mudstone [5076 7379]. Richardson 1928, pp.95–96.

REFERENCES

BUTLER, A. J. 1946.   Water supply from underground sources of the Birmingham-Gloucester district.   *Geol. Surv. GB Wartime Pamphlet* No. 32 Part III.

CROFTS, R. G. 1982.   The sand and gravel resources of the country between Coventry and Rugby, Warwickshire: description of 1:25 000 sheets SP 47 and part of SP 37.   *Miner. Assess. Rep. Inst. Geol. Sci.*, No. 125.

INSTITUTE OF GEOLOGICAL SCIENCES. 1978.   IGS Boreholes 1976.   *Rep. Inst. Geol. Sci.* No. 77/10.

— 1979.   IGS Boreholes 1978.   *Rep. Inst. Geol. Sci.* No. 79/12.

— 1980.   IGS Boreholes 1979.   *Rep. Inst. Geol. Sci.* No. 80/11.

RICHARDSON, L. 1928.   Wells and Springs of Warwickshire. *Mem. Geol. Surv. GB.*

APPENDIX 2

List of Geological Survey photographs

Copies of these photographs may be seen in the Library of the British Geological Survey, Keyworth, Nottingham NG12 5GG. Prints and lantern slides may be bought. National Grid References, all in 100-km square SP, are those of the viewpoints. The photographs belong to Series A.

Permian

13093	Load casts in sandstone, Ashow Formation, Whitemoor Brickworks, Kenilworth [2956 7180] (Plate 3)
13094	Sandstone, Ashow Formation, Enville Group, Whitemoor Brickworks, Kenilworth [2956 7180]
13097	Sandstone, Ashow Formation, Enville Group, Cherry Orchard Brickworks, Kenilworth [2960 7213]
13098	Kenilworth Castle from the south. The castle is constructed from locally obtained Enville Group sandstone (Kenilworth Sandstone Formation) [2795 7203]
13099	Breccia in wall of Kenilworth Castle [2785 7220]
13100	Breccia in Kenilworth Sandstone Formation, Castle Quarry, Kenilworth [2779 7194] (Plate 2)
13101	Flaggy sandstone, Kenilworth Sandstone Formation, Castle Quarry, Kenilworth [2773 7193]
13102	Gibbet Hill Conglomerate, Gibbet Hill, opposite the entrance to Warwick University [3045 7522]
13104	Stoneleigh Church [3300 7357]

Triassic

13066	Bromsgrove Sandstone, cutting on A444 road, Blackdown [3175 6875]
13067	Channel lag conglomerate and cross-bedded sandstone, Bromsgrove Sandstone, Guy's Cliffe, Warwick [2937 6678] (Plate 4)
13068	Sandstones showing trough cross-bedding, Bromsgrove Sandstone, Guy's Cave, Guy's Cliffe, Warwick [2934 6679]
13092	Warwick Castle [2878 6494] (Plate 1)
13105, 13106	Cross-bedded sandstone, Bromsgrove Sandstone, small quarry on west side of A444 road [2897 6914]
13107, 13108	Cross-bedded intraformational sandy conglomerates and sandstones, Bromsgrove Sandstone, old quarry, Woodloes Estate [2798 6654]
13109	Bromsgrove Sandstone, Coton End Quarry, Warwick [2899 6551]
13113	Massive sandstone overlying planar-bedded sandstone disturbed by dewatering, Bromsgrove Sandstone, Blackdown Hill [3231 6885] (Plate 5)
13114	Sandstone and mudstone, Bromsgrove Sandstone, old quarry, Blackdown [3230 6886]

Jurassic

8511–15	Marlstone Rock Bed and Middle Lias silts and clays, Napton on the Hill Brickworks [455 613]
13069, 13070	Blue Lias, Long Itchington Quarry [419 629]
13071, 13072	Blue Lias, Stockton Quarry [442 641]
13077	Lower Lias clays overlying Blue Lias, Harbury Quarry [387 588]
13074	Harbury Church, constructed mostly of local Blue Lias and Langport Member limestones [374 600]
13079–81	Blue Lias, Rugby Quarry [4950 7589; 4950 7595] (13079 Plate 6)
13082, 13083	Anticlines in Blue Lias, Rugby Quarry [4940 7599; 4948 7595]
13084	Blue Lias, Newbold Quarry, Rugby [4965 7704]
13103	Southam Church, contructed of locally quarried Blue Lias and Langport Member limestones [418 617]

Quaternary

8540–42	Boulder Clay overlying current-bedded glacial sand, Hillmorton Sand Pit, junction of Lower Rugby Road with Featherbed Lane [532 745]
9346	Roman embankment resting on alluvium at 0.6 m below present alluvium level, Roman camp, Fosse Way, Chesterton (view point not known)
9345	Alluvium and 1st and 4th terraces of the River Avon, Barford-Wasperton road (view point not known)
13073	Landslips in Penarth Group escarpment, Ufton [373 625]
13075	Glacial drift, Harbury Quarry [3822 5884]
13085	Thrussington Till overlying cross-bedded Baginton Sand and Gravel, Wolston Pit [4098 7466] (Plate 7)
13086	Wolston Sand, Lower Wolston Clay and Baginton Sand and Gravel, Wolston Pit [4104 7463]
13087–88	Lower Wolston Clay and Thrussington Till overlying Baginton Sand and Gravel, Ryton Quarry [377 727]
13089	Lower Wolston Clay overlying Thrussington Till, Ryton Quarry [3763 7261]
13090–91	Thrussington Till overlying Baginton Sand and Gravel, Manor Farm Gravel Pit, Ryton-on-Dunsmore [3919 7357]
13095	Drift-filled channel in Enville Mudstone, Cherry Orchard Brickworks, Kenilworth [295 722]
13096	Baginton Sand and Gravel, Cherry Orchard Brickworks, Kenilworth [2945 7226]
13110	Baginton Sand and Gravel, Church Farm Gravel Pit, Ryton on Dunsmore [3816 7487]
13111	Lower Wolston Clay, Thrussington Till and Baginton Sand, Brandon Wood Gravel Pit, Brandon [3871 7643]
13112	Cross-bedded Baginton Sand and Gravel, Brandon Wood Gravel Pit, Brandon [3871 7643]

APPENDIX 3

Principal sections of Jurassic rocks

In the following sections, depths within the beds whose thicknesses are given in the right hand column are measured from the tops of those beds unless otherwise stated.

Harbury Quarry (composite section) [385 589]

	Thickness m
LOWER LIAS	
Mudstone; pale grey, blocky to fissile; pyrite nodules and bands	0.50
Limestone; pale cream to green-grey, rubbly; passing to	0.30
Mudstone; medium to dark grey; blocky in upper part, very fissile below with pyrite	0.55
Limestone; pale grey, weathering to buff, very fine-grained, earthy; small iron-stained burrows	0.27
Mudstone, pale grey, blocky; shells including *Gryphaea*; passing to	0.24
Paper shale; grey, with iron-stained lenses, shells and shell debris; burrows?; irregular 0.02 m band of pale grey mudstone at 0.13 m; dark grey at base with pyritic nodules and lenses	0.31
Mudstone; pale grey, blocky; harder nodules and scattered shells	0.35
Cementstone; grey, very fine-grained, partially crystalline; iron-stained burrows; scattered shell debris; buff, iron-stained weathered surfaces	0.17
Mudstone; pale grey, blocky to poorly fissile; shells, pyritic nodules, selenite, burrows, wood debris; impersistent pyrite bed at 0.1 m	c.1.00
Obscured by scree; presumed mudstone and paper shale	c.7.00
Mudstone; grey, poorly fissile; ammonites, shell debris, burrows, some selenite	c.1.00
BLUE LIAS	
Cementstone (Top Rock); dark grey, poorly fissile; earthy in top 0.05 m with *Angulaticeras sp.*, and *Vermiceras scylla* (Reynès); scattered shells and small pyritic trails	0.37
Mudstone; grey, poorly fissile; selenite, shells and shell detritus, burrows; pyrite nodules concentrated in layer at 0.22 m, mudstone more blocky below	0.35
Cementstone; grey, very fine-grained; iron-stained burrows and joints; sporadic shells	0.13
Mudstone; grey, blocky to poorly fissile; shells, spat, selenite, iron-stained lenses and patches, burrows	0.40
Cementstone; grey, partly crystalline, very fine-grained; scattered shells	0.18
Mudstone; grey, blocky to poorly fissile; hard and calcareous to 0.15 m; shells and shell detritus, burrows, some selenite; iron-staining along vertical and horizontal joints, 0.06 m grey, nodular cementstone at 0.21 m with a few shells; darker grey towards base	0.31
Cementstone; grey; scattered shells	0.14
Mudstone; grey; fissile at top; pyritised burrows, selenite, pyrite nodules; some bedding planes crowded with shells, mainly bivalves	0.30

	Thickness m
Paper shale, dark grey, pyritic; selenite, burrows, pyrite nodules, bands of shell debris	0.11
Mudstone, grey, fissile; pyrite nodules	0.05
Cementstone, grey; scattered shells, iron-stained burrows	0.12
Mudstone, grey, blocky to poorly fissile; shells, burrows, selenite; 0.03 m pyrite band at 0.20 m	0.25
Cementstone; medium to dark grey; scattered shells, pyritised burrows	0.17
Mudstone; grey, blocky; bioturbated to 0.15 m; shells and shell debris, scattered selenite, burrows, *Chondrites*; poorly fissile from 0.50 to 0.60 m; harder and more calcareous in basal 0.10 m	0.77
Cementstone; grey, nodular; scattered shells, burrows	0-0.15
Mudstone; grey, blocky, shelly; fissility, selenite and pyrite content increase towards base; very dark grey to 0.08 m; basal 0.10 m highly calcareous with many pyrite nodules and burrows	0.36
Cementstone; grey; scattered horizontal burrows, shells	0.19
Mudstone; pale to medium grey, blocky to poorly fissile, hard and calcareous; scattered pyritised burrows	0.20-0.23
Cementstone; grey, partly crystalline, nodular	0.0-0.09
Paper shale; dark grey, with selenite	0.06
Mudstone, grey, blocky, calcareous	0.02
Cementstone, grey; scattered shells, pockets of shell debris, veins of calcite	0.22
Mudstone, pale to dark grey, fissile; many burrows, *Chondrites*	0.08
Mudstone; pale grey, blocky, calcareous; burrows, shells and shell debris, scattered selenite	0.23
Paper shale; dark grey; many burrows in top 0.05 m, commonly filled with pale grey material; selenite on bedding planes; shells very common on some bedding planes; woody fragments	0.15
Mudstone; pale grey, blocky, calcareous, bioturbated; shells and shell detritus, scattered pyrite specks; poorly fissile below 0.10 m and less bioturbated with more shells	0.18
Paper shale; dark grey; *Chondrites* in top 0.02 m, selenite and shell detritus common on bedding planes	0.13
Mudstone; pale grey, blocky; shells and shell detritus, burrows, scattered selenite	0.14
Mudstone; grey, poorly fissile; passing to	0.18
Paper shale; dark grey; much selenite, shells and shell detritus	0.08
Mudstone; pale grey, blocky; shells and shell detritus, pyrite nodules	0.11
Paper shale; pale grey; *Chondrites* to 0.05 m; selenite, shells and shell debris; grey-brown in basal 0.09 m; passing to	0.16
Mudstone; pale grey, blocky; scattered burrows, shells and shell detritus	0.03
Cementstone; grey, partly crystalline; scattered shells	0.24
Mudstone; grey, blocky; shells, shell detritus, sporadic burrows, scattered selenite, scattered pyrite specks; 0.03 m band at 0.17 m of more fissile mudstone with much shell debris	0.26
Cementstone; grey, partly crystalline; scattered shells	0.23
Mudstone; grey, blocky; shells and shell detritus; calcareous to 0.8 m, poorly fissile below	0.12
Paper shale; dark grey; pyritised burrows,	

BLUE LIAS (cont.)

	Thickness m
Chondrites, scattered shells and shell detritus; pyrite nodules commoner towards base	0.16
Mudstone, pale to medium grey, blocky; shells, burrows, spheroidal weathering in parts	0.25
Cementstone, grey, partly crystalline; partly pyritised wood fragment	0.13
Mudstone, pale grey, blocky to poorly fissile; shells, shell detritus, selenite, burrows, pyrite nodules, *Gryphaea*	up to 0.34
Cementstone, grey; scattered shells, pyrite specks; passes laterally into calcareous mudstone with burrows	0.0 – 0.12
Mudstone; grey, blocky; calcareous at base	0.23
Cementstone, grey; passes laterally into calcareous mudstone	0.0 – 0.11
Mudstone, grey, blocky, calcareous	0.06
Cementstone, grey; scattered shells, pyrite specks and woody fragments; irregular top and base	0.19 – 0.40
Mudstone; grey, blocky to poorly fissile; burrows, selenite	0.16
Cementstone, pale to medium grey; scattered shells	0.10
Mudstone, grey, hard, calcareous, blocky; shells and shell detritus, slickensided joints, slightly more fissile at base	0.32
Cementstone, grey, pyritised joints, scattered shell debris	0.16
Mudstone; grey, blocky at top, calcareous, slightly more fissile to base	0.10
Cementstone, lenticular; pyritic along joints; passes laterally into calcareous mudstone; *Liostrea*, *Gryphaea*	0 – 0.12
Mudstone, grey, blocky; fissile locally and near base; selenite common in fissile areas; pyrite, shells; extensively burrowed at base	0.32
Cementstone (Thick Rock); grey; pyrite specks, slickensided joints, scattered shell debris, *Liostrea*	0.47
Mudstone; medium to dark grey, poorly fissile to blocky, highly bioturbated; selenite on some bedding planes, shells and shell detritus	0.45
Cementstone; grey, crystalline; U-shaped burrows, scattered rhynchonellids	0.07
Mudstone; medium to dark grey, poorly fissile; shells and shell detritus	0.05
Cementstone; medium to dark grey; horizontal and vertical burrows; irregular top surface, possibly bored	0.15
Mudstone; grey, shelly, poorly fissile; pyrite specks; impersistent nodular cementstone	0.18
Cementstone; grey, partly crystalline; scattered shells	0.15
Mudstone	0.15
Cementstone	0.10

Stockton Quarry [440 640]

LOWER LIAS

Cementstone; grey, weathered to buff	0.10
Mudstone; weathered, some paper shale	2.20

BLUE LIAS

Cementstone (Top Rock); pale grey; scattered shell debris; ammonites common in top 0.05 m, mainly *Vermiceras scylla* (Reynès)	0.41
Mudstone; grey, fairly fissile, marly	0.15
Paper shale; dark grey; selenite on bedding planes, shells and shell debris, spat	0.25
Mudstone, pale grey, fissile	0.20
Paper shale, medium to dark grey; abundant shell debris, spat, pectinids, echinoid spines, selenite on bedding planes; pyritic in basal 0.05 m; marly in basal 0.01 m	0.25
Cementstone, pale grey; scattered shell debris; irregular base and top	0.21
Mudstone, grey, slightly blocky; crushed bivalves	0.25
Cementstone; grey, very earthy; abundant shell debris	0.09
Mudstone; dark grey, fissile; fissility increases towards base; crushed shell debris, pectinids, spat, iron staining; rare cementstone nodules in lowest 0.3 m	0.42
Paper shale; dark grey to black; pyrite layer 0.04 m from base; abundant selenite and shells	0.23
Cementstone; pale grey, partly crystalline; scattered shell debris	0.18
Mudstone, grey; shell debris, spat; marly at top, increasingly fissile towards base	0.15
Paper shale; black; abundant selenite, shells and debris including *Gryphaea*	0.15
Cementstone; pale grey, shelly, partly crystalline; shell band at 0.11 m	0.18
Mudstone; grey, pyritic; marly in top 0.07 m; more fissile and darker grey towards base; shells and selenite commoner towards base	0.82
Cementstone; pale grey, partly crystalline; scattered shell debris	0.22
Mudstone; grey and blocky at top, more fissile and darker grey towards base; pale grey nodular cementstone up to 0.08 m thick with shell debris	0.20
Paper shale; dark grey, pyritic	0.03
Cementstone; pale grey	0.11
Mudstone; grey, blocky to fissile; selenite; 0.06 m nodular cementstone at base	0.23
Mudstone; dark grey, slightly fissile; sporadic shell debris	0.11
Paper shale, black; abundant selenite and spat	0.22
Mudstone, dark grey, fairly fissile	0.17
Paper shale, black; abundant selenite, spat and shell debris	0.17
Cementstone; pale grey, partly crystalline; scattered shell debris, rare *Calcirhynchia calcaria*	0.13
Mudstone, dark grey, fissile; scattered shell debris	0.10
Paper shale, black; abundant selenite, sporadic shell debris	0.12
Mudstone, dark grey, fissile; sporadic shell debris	0.13
Paper shale, black; much selenite, spat and shell debris; more blocky in basal 0.05 m	0.22
Cementstone, pale grey, partly crystalline; scattered shell debris	0.23
Mudstone, pale grey, marly; blocky at top, more fissile and darker grey towards base	0.28
Cementstone, pale grey; rare shell debris	0.23
Mudstone, dark grey, blocky to poorly fissile; pyrite-rich band at base	0.21
Paper shale, black; abundant selenite and shell debris	0.15
Mudstone, grey, blocky; horizontal pyrite traces (burrows); marly at base; passing to	0.31
Cementstone, pale grey; rare shell debris; passing to	0.14
Marl, pale grey; ?calcified hard in places; passing to	0.12
Mudstone, dark grey and fissile at top, paler grey, marly and blocky at base; shell debris, selenite	0.26
Cementstone, pale grey; rare shell debris	0.14

BLUE LIAS (cont.)

	Thickness m
Mudstone, medium to dark grey, marly; fissile at top with shell debris and selenite; nodular cementstone up to 0.1 m thick at 0.2 m	0.37
Mudstone, pale to dark grey, blocky; scattered selenite and spat	0.19
Limestone, pale grey, finely crystalline; comminuted shell debris	0.11
Mudstone, pale to dark grey, blocky; selenite, rare shell debris	0.42
Cementstone, dark grey, partly crystalline	0.19
Mudstone, dark grey, blocky; passing to	0.14
Cementstone; pale grey, earthy, nodular; commonly passes into	0.13
Mudstone; pale grey, blocky to poorly fissile; marly at base, pyritic above 0.01 m shale layer	0.42
Cementstone (Thick Rock); pale grey; passing to	0.38
Mudstone; pale grey, bioturbated, blocky to poorly fissile	0.23
Cementstone; pale grey, partly crystalline; shell debris and shells, partly developed burrows, *Diplocraterion*	0.06 – 0.13
Mudstone, pale grey, poorly fissile; shell debris, spat	0.03
Cementstone; pale grey	0.10
Mudstone; pale grey, slightly fissile; nodular cementstones up to 0.5 thick at 1.3 m; more fissile beneath nodules	0.16
Cementstone; pale grey, poorly crystalline; shell debris	0.14
Mudstone; pale grey, blocky to fissile; shell debris	0.08
Cementstone; pale grey; very thin calcite veins	0.14
Mudstone; pale grey; increasingly fissile towards base; pyritic traces, scattered shell debris	0.20
Paper shale; dark grey, pyritic; abundant selenite, *Chondrites*, shell debris	0.04
Mudstone; grey, fissile; marly at base	0.08
Cementstone; pale grey; pyritic traces, sub-vertical burrows	0.14
Mudstone; grey; blocky at top, increasingly fissile towards base; passes laterally to paper shale	0.29
Cementstone; pale grey, nodular; shell detritus	up to 0.08
Cementstone; pale grey; shell debris, horizontal burrows or traces at top, *Calcirhynchia calcaria* abundant in places, large bivalves, *Gryphaea*, *Plagiostoma giganteum*	0.16
Mudstone; pale to medium grey, blocky	0.03 – 0.07
Cementstone; pale grey; scattered comminuted shell debris, thin calcite veins, *Gryphaea*	0.10
Mudstone; pale grey, blocky; more fissile towards base	0.04
Paper shale; dark grey to black; abundant selenite on bedding planes, scattered shell debris and spat; pyritic lenses common at 0.07 m and 0.17 m	0.22
Mudstone; pale grey, blocky	0.07 – 0.09
Cementstone; pale grey, nodular; scattered shell debris	up to 0.09
Mudstone; grey to dark grey, slightly fissile; abundant shell debris, *Plagiostoma*, *Gryphaea*; pyrite band at 0.09 m	0.17
Cementstone; pale to medium grey, partly crystalline; sporadic shell debris, irregular sub-vertical burrows c.0.01 m wide and filled with darker grey material	0.10
Mudstone; grey, blocky; horizontal pyrite traces	0.12
Cementstone; pale grey; short blackened micro-joints	0.22

Mudstone; grey; slightly fissile to 0.15 m and from 0.4 m to base; central part blocky; increasingly pyritous and iron-stained downwards	0.53
Cementstone (Worm Bed); pale grey; scattered shell debris; vertical burrows up to 0.015 m wide transect whole bed and branch irregularly in places	0.13
Mudstone, dark grey; blocky at top, fissile at base; iron staining common towards base	0.14
Cementstone; pale grey; sporadic shell debris, pyritic traces, *Spiriferina*	0.07
Mudstone; dark grey, fissile	0.09
Paper shale; dark grey; abundant selenite on bedding planes, crushed bivalves, small pectinids, spat, rare ammonite impressions, *Gryphaea*	0.13
Mudstone; dark grey, fairly fissile	0.11
Cementstone; pale to medium grey, partly crystalline; rare shell debris	0.16
Mudstone; grey; blocky at top, fissile at base; woody fragment c.0.05 m long and iron-stained on outer surface	0.20
Cementstone; grey; rare shell debris	0.06
Mudstone (below water level)	0.2

Old Quarry, Stockton [438 649]

BLUE LIAS

Clay	0.60
Cementstone; grey with ochreous patches	0.18
Clay; yellow-grey (weathered mudstone)	0.36
Cementstone; grey with ochreous patches; vertical burrows, pyritic traces, shell debris	0.14
Clay; grey-yellow; blocky to 0.3 m, fissile below; flattened shells and shell debris; pyritic band at 0.45 m	0.50
Cementstone; grey weathering to buff; thin calcite veins	0.15
Mudstone; grey to dark grey, marly; slightly fissile to 0.24 m, very fissile and dark grey below with pyrite band	0.30
Cementstone; grey; much shell debris, scattered rhynchonellids; irregular base	0.16
Mudstone; pale to medium grey; marly and blocky to 0.10 m, fissile with shell debris and spat below; iron-staining on some bedding planes; pyritic lenses near base; 0.04 m cementstone nodule at 0.7 m; *Liostrea*	0.91
Cementstone; dark grey; rare shell debris; burrows at base	0.14
Mudstone, dark grey, blocky	0.13
Cementstone; grey, nodular; scattered shell debris	0.08
Mudstone, dark grey, fissile, pyritic	0.08
Cementstone, pale grey; shell debris; burrowed top surface	0.16
Mudstone, pale grey, blocky, marly; compacted in basal 0.05 m	0.22
Paper shale, dark grey to black; abundant selenite and shell debris, scattered pyrite	0.20
Mudstone; dark grey, fissile; abundant shells and shell debris, selenite	0.21
Paper shale; dark grey to dark brown; abundant shells, mainly thin crushed bivalves, abundant selenite, ammonite fragments, fish debris, *Chondrites*	0.12
Mudstone; dark grey, very fissile; marly at base	0.08
Cementstone; grey; uncompacted and marly in places	0.08
Paper shale; dark brown-grey; abundant shells, mainly spat and thin bivalves, some fish debris,	

	Thickness
BLUE LIAS (cont.)	m
Chondrites, ammonite fragments, selenite	0.22
Mudstone; pale grey, poorly fissile; shell debris, spat, *Chondrites*	0.12
Paper shale; dark grey-brown; abundant *Chondrites*, crushed bivalves (pectinids) and spat, ammonite fragments, selenite	0.17
Mudstone; pale grey, fairly fissile, marly	0.07
Cementstone; grey, yellow-weathering; shell debris, scattered shells, *Gryphaea*	0.25
Mudstone; grey, fissile to very fissile; woody fragments with iron-stained surface	0.12
Paper shale; pale to medium grey; abundant selenite, clusters of shells and shell debris	0.15
Cementstone; pale grey; irregular top infilled with mudstone, scattered shell debris, subhorizontal burrows	0.19 – 0.26
Mudstone, pale grey, slightly fissile, marly; passing to	0.12
Paper shale, dark brown to black; shells and shell debris, spat, ostracods, abundant selenite	0.25
Mudstone, pale grey, slightly fissile to blocky; marly at base; patches of selenite, scattered shell debris	0.27
Cementstone, pale to medium grey; blocky to 0.5 m, darker grey and more fissile to 0.55 m, marly to base	0.60
Cementstone; pale grey; sporadic recrystallised shell debris; base not seen	0.10

APPENDIX 4

Sources of geophysical data

GRAVITY DATA

The Bouguer anomaly map (Figure 21) is based on investigations by the British Geological Survey between 1954 and 1978. The anomaly values are referred to the 1967 International Gravity Formula and to the 1973 National Gravity Reference Net (Masson Smith and others, 1973). The contour map uses values calculated at a constant density of 2.4 t/m^3; since most of the land lies between 60 m and 100 m OD and average near-surface density values rarely lie outside the range 2.3–2.5 t/m^3, any errors due to this assumption should be less than 0.5 mGal. The anomaly values include a correction, usually less than 0.1 mGal, for the effect of topography.

Working in this same general area, Cook and others (1952) deduced a density contrast between the Carboniferous and older rocks of 0.23 t/m^3. Parasnis (1952) obtained mean densities of 2.48 t/m^3 for the Coal Measures and 2.71 t/m^3 for the underlying rocks. The density values used in the present interpretation are based on laboratory measurements on rock samples, formation density borehole logs, and measurements of the gravitational field at different depths underground, the last being the most satisfactory way of determining representative values for large volumes of rock. Only the Carboniferous rocks have been extensively sampled (Parasnis, 1952; Cornwell *in* Poole, 1977, 1978); an average density of 2.49 ± 0.02 t/m^3, increasing with depth, was applicable over a thickness of about 500 m. Values for the Mesozoic rocks vary over the wider range of 2.2–2.5 t/m^3, depending upon lithology and degree of compaction, but 2.4 t/m^3 is regarded as being representative. The Cambrian sediments have been assigned a density of 2.7 t/m^3. Densities and values for any other Lower Palaeozoic sediments are expected to lie within the range 2.68–2.72 t/m^3 for the more acid intrusives (Evans and Maroof, 1976) but may increase to over 2.8 t/m^3 for intermediate and basic types (Poole, 1977, 1978). Within the Precambrian, densities probably range from 2.65 t/m^3 (acidic) to over 3 t/m^3 (basic).

MAGNETIC DATA

Total magnetic field anomalies shown in Figure 22 have been taken from the original work-sheets in order to preserve the maximum amount of detail in the contouring. The values were adjusted for a gradient taken from a generalised description of the geomagnetic field which differs from the local reference field adopted subsequently for the British Isles. The latter, which increases in the directions of grid north and grid west at the unform rates of 2.1728 nT/km and 0.259 nT/km respectively, has been used in preparing the published 1:250 000 scale magnetic anomaly map series. Data were collected along west–east flight lines 1.6 km apart at a constant height of 550 m above sea-level. Although the amplitude of the field varies by more than 350 nT there are fewer intense, localised anomalies than in areas to north and south, and this is consistent with the low magnetisation expected from the sediments that are known to extend to a least 1000 m depth. No magnetic properties have been determined for the Warwick district but relevant values are available from around Hinckley (Cornwell and Allsop, 1981). Susceptibilities for the sediments are less than 0.2×10^{-3} SI in the Triassic and Carboniferous, increasing to about 0.7×10^{-3} SI in the Cambrian and Precambrian strata; the igneous rocks such as granodiorite and syenite gave significantly higher values of up to 25×10^{-3} SI although they ranged widely; even allowing for remanent magnetisation their average apparent susceptibility is probably closer to 5×10^{-3} SI.

SEISMIC DATA

Geophysical logging of the many NCB boreholes has proved particularly useful in identifying and correlating formations, and in defining the properties (such as porosity) of specific horizons. In addition there have been extensive seismic reflection surveys which have shown seismic velocities to increase from about 2.5 km/s near the surface to 4 km/s towards the base of the Carboniferous. It is not usually possible to identify the base of the Mesozoic rocks. The discontinuous nature of reflecting horizons within the Carboniferous succession may be attributable to lateral facies changes, but strong persistent reflections were recorded from near its base where the thicker coal seams occur. Little information is available from below that level.

FOSSIL INDEX

No distinction is made here between a positively determined genus or species and examples doubtfully referred to them (i.e. with the qualifications aff., cf. or ?).

GENERAL INDEX

Page numbers in italics refer to illustrations

BRITISH GEOLOGICAL SURVEY

Keyworth, Nottingham NG12 5GG

Murchison House, West Mains Road,
Edinburgh EH9 3LA

The full range of Survey publications is available
through the Sales Desks at Keyworth and
Murchison House. Selected items are stocked by
the Geological Museum Bookshop, Exhibition
Road, London SW7 2DE; all other items may be
obtained through the BGS London Information
Office in the Geological Museum. All the books
are listed in HMSO's Sectional List 45. Maps are
listed in the BGS Map Catalogue and Ordnance
Survey's Trade Catalogue. They can be bought
from Ordnance Survey Agents as well as from
BGS.

*The British Geological Survey carries out the geological
survey of Great Britain and Northern Ireland (the latter as
an agency service for the government of Northern Ireland),
and of the surrounding continental shelf, as well as its
basic research projects. It also undertakes programmes of
British technical aid in geology in developing countries as
arranged by the Overseas Development Administration.*

*The British Geological Survey is a component body of the
Natural Environment Research Council.*

Maps and diagrams in this book use topography
based on Ordnance Survey mapping

HER MAJESTY'S STATIONERY OFFICE

HMSO publications are available from:

HMSO Publications Centre
(Mail and telephone orders)
PO Box 276, London SW8 5DT
Telephone orders (01) 622 3316
General enquiries (01) 211 5656
Queueing system in operation for both numbers

HMSO Bookshops
49 High Holborn, London WC1V 6HB
   (01) 211 5656 (Counter service only)
258 Broad Street, Birmingham B1 2HE
   (021) 643 3740
Southey House, 33 Wine Street, Bristol BS1 2BQ
   (0272) 264306
9 Princess Street, Manchester M60 8AS
   (061) 834 7201
80 Chichester Street, Belfast BT1 4JY
   (0232) 238451
71 Lothian Road, Edinburgh EH3 9AZ
   (031) 228 4181

HMSO's Accredited Agents
(see Yellow Pages)

And through good booksellers